A Practical Guide
to the Forensic
Examination of Hair

A Practical Guide to the Forensic Examination of Hair

From Crime Scene to Court

James Robertson and
Elizabeth Brooks

CRC Press
Taylor & Francis Group
Boca Raton London New York

CRC Press is an imprint of the
Taylor & Francis Group, an **informa** business

CRC Press
Boca Raton and London

First edition published 2022

by CRC Press
6000 Broken Sound Parkway NW, Suite 300, Boca Raton, FL 33487-2742

and by CRC Press
2 Park Square, Milton Park, Abingdon, Oxon, OX14 4RN

ISBN: 978-1-138-62861-8 (hbk)
ISBN: 978-1-032-10861-2 (pbk)
ISBN: 978-1-3152-1065-0 (ebk)

DOI: 10.4324/9781315210650

Typeset in Sabon
by KnowledgeWorks Global Ltd.

Contents

Preface   xi

About the Authors   xiii

Chapter 1 **Historical Context to Contemporary**
           Practice in Hair Examinations   **1**

           1.1   Introduction   1

           1.2   A Brief History of the Last 100 Years
                  of Hair Examination   4

           1.3   The Criminalistics Value of Hairs in the
                  Context of Contemporary Case Management   12

           1.4   Implications Arising from the Review into
                  the FBI Approach to Hair Reporting   18

           1.5   Where to Now—Final Conclusions and
                  the Way Forward   22

Chapter 2 **Fundamentals of the Biology and Chemistry**
           of Hairs   **25**

           2.1   Introduction   25

           2.2   How is the First Hair Follicle Formed?   27

           2.3   The Role of the Hair Follicle in the Growth
                  of Hair Fibres   29

           2.4   The Hair Cycle   32

           2.5   Hair Distribution, Types and Growth Rates   36

                  2.5.1   Distribution   36

                  2.5.2   Types   37

                  2.5.3   Growth Rates   37

2.6	Morphology and Anatomy		38
	2.6.1	Morphology	38
	2.6.2	Anatomy	39
		2.6.2.1 Cuticle	41
		2.6.2.2 Cortex	43
		2.6.2.3 Medulla	45
2.7	Hair Colour and Pigmentation		46
	2.7.1	Visual and Macroscopic Assessment of Natural Hair Colour	46
	2.7.2	Pigment Basis for Hair Colour	47

Chapter 3 Recognition, Recording and Recovery Considerations — **49**

3.1	Introduction		49
3.2	Recording and Recovery of Hairs		50
	3.2.1	Scene Considerations	50
	3.2.2	Scene Sampling Protocols	51
	3.2.3	Sampling from Deceased Persons or Remains	53
	3.2.4	Sampling from Living Persons	55
3.4	Conclusions		56
	Appendix 3.1: Forensic Known Hair Collection Kit		57

Chapter 4 Laboratory Examinations — **61**

4.1	Introduction		61
4.2	Level 1 Examinations—Recognition and Separation of Human and Animal Hairs		63
4.3	Level 2 Examinations—Examination of Human Hairs for Body Area and Ethnic Origin and Selection for DNA Analysis		69
	4.3.1	Body Area Determination	69
	4.3.2	Ethnic Origin	71
	4.3.3	Selection of Hairs for DNA Analysis	72
	4.3.4	Low-Power Microscopic Examination of Hairs	76

4.4   Level 3 Examinations—Detailed Examination
       of Hairs and Comparison Microscopy   84

       4.4.1   High-Power Transmitted Light
                Microscopic Examination   92

       4.4.2   Comparison Microscopy   110

Appendix 4.1A: Animal Hair Examination Proforma   114

Appendix 4.1B: Animal Hair Examination Proforma   115

Appendix 4.2A: Human Hair Examination Proforma,
     LPM   116

Appendix 4.2B: Human Hair Examination Proforma,
     TLM   117

Appendix 4.2C: Human Hair Examination Proforma
     Comparison   118

Appendix 4.3: Features of Animal Hairs   119

Appendix 4.4: Nuclear Staining of Telogen Roots   120

Appendix 4.5: Test for Bleached Hair   126

Chapter 5   Evaluation and Interpretation   127

5.1   Introduction   127

5.2   Transforming Data into Information   127

5.3   Formulating an Opinion   132

5.4   Estimating Probabilities   132

5.5   Defining Error   133

Chapter 6   Reporting   139

6.1   Scope   139

6.2   Report Formats   140

6.3   Issuing of Reports   141

6.4   Report Contents   142

       6.4.1   General Requirements   142

       6.4.2   Analysis and Comparison of Material   143

       6.4.3   Reporting Conclusions and Opinions   144

       6.4.4   Wording Used in Reporting Hair
                Conclusions   145

       6.4.5   The Role of Statistics and Verbal
                Scales for Hair Opinions   148

| | 6.4.6 | Dealing with Cognitive Bias | 152 |
| | 6.4.7 | Testimony and Giving Evidence | 153 |

Appendix 6.1: Protocol for Hair Examination   157

Appendix 6.2: Example Report 1   158
    1. Custody of Items   159
    2. Examination and Results   159
    3. Conclusions   160
    4. Hair Appendix   161
    5. References   162

Appendix 6.3: Example Report 2   162
    1. Custody of Items   163
    2. Examination and Results   163
    3. Conclusions   164
    4. Hair Appendix   165
    5. References   166

Appendix 6.4: Example Report 3   166
    1. Custody of Items   167
    2. Examination and Results   167
    3. Conclusions   168
    4. Hair Appendix   168
    5. References   169

Chapter 7   Training Considerations   171

7.1   Scope   171

7.2   Assumed Knowledge and Competencies   172

7.3   Level 1 Training—Recognition and Separation
      of Human and Animal Hairs   173

    7.3.1   Overview   173

    7.3.2   Module 1   173
        7.3.2.1   Learning Outcomes   173
        7.3.2.2   Content   174
        7.3.2.3   Competency Evaluation 174

    7.3.3   Module 2   174
        7.3.3.1   Learning Outcomes   174

Contents

		7.3.3.2	Content	175
		7.3.3.3	Competency Evaluation	175
	7.3.4	Module 3		175
		7.3.4.1	Learning Outcomes	175
		7.3.4.2	Content	175
		7.3.4.3	Competency Evaluation	176
	7.3.5	Module 4		176
		7.3.5.1	Learning Outcomes	176
		7.3.5.2	Content	176
		7.3.5.3	Competency Evaluation	177
7.4	Level 2 Training—Examination of Human Hair Including Body Area and Ethnic Origin and Selection of Hairs for DNA Testing			177
	7.4.1	Overview		177
	7.4.2	Module 1		177
		7.4.2.1	Learning Outcomes	177
		7.4.2.2	Content	178
		7.4.2.3	Competency Evaluation	178
	7.4.3	Module 2		178
		7.4.3.1	Learning Outcomes	178
		7.4.3.2	Content	178
		7.4.3.3	Competency Evaluation	179
	7.4.4	Module 3		179
		7.4.4.1	Content	179
		7.4.4.2	Competency Evaluation	180
7.5	Level 3 Training—Detailed Examination of Hairs and Comparison Microscopy			180
	7.5.1	Overview		180
	7.5.2	Module 1		181
		7.5.2.1	Content	182
		7.5.2.2	Competency Evaluation	182
	7.5.3	Module 2		182
		7.5.3.1	Content	182
		7.5.3.2	Competency Evaluation	183

	7.5.4	Module 3	183
		7.5.4.1	Content	184
		7.5.4.2	Competency Evaluation	184
	7.5.5	Module 4	185
		7.5.5.1	Content	186
		7.5.5.2	Competency Evaluation	186
	7.4.6	Module 5	187
		7.4.6.1	Content	187
		7.4.6.2	Competency Evaluation	187

Chapter 8 **Acquired Characteristics**   **189**
Adine Boehme (Guest author)

8.1	Introduction	189
8.2	Examinations with Acquired Characteristics	190
8.3	Reporting Acquired Characteristics	191
8.4	Natural and Unnatural Hair Loss	192
8.5	Cut Hairs—Professional Hairdressing	192
8.6	Motor Vehicle Collision and Windscreen Impact	193
8.7	Blunt-Bladed Implement	195
8.8	Sharp-Bladed Implement—Kitchen Knife	196
8.9	Blunt Force Impact—Broken/Crushed Hair	198
8.10	Blunt Force Impact—Floor Safe Door	199
8.11	Arson/Fire Versus Cosmetic Heat Styling	200
8.12	Arson/Explosion	201
8.13	Explosion	203
8.14	Head Lice	205
8.15	Conclusion	208

Glossary	209
References	215
Index	227

Preface

The evolution and development of DNA analysis on the practice of forensic science has been a dominant factor in the last 30 years, placing a very heavy focus on its value in identifying the source of biological materials. Although there is a greater appreciation of the need to evaluate and understand how recovered biological materials relate to what is alleged to have taken place, and not simply the DNA "number" and individualisation, nonetheless it can be argued with some justification that the impact of DNA has led to an emphasis on the value of other forms of forensic evidence in a narrow identification paradigm. DNA has also been held up as the "gold standard" that other forms of forensic endeavour must aspire to meet. In this context, hairs have been the subject of critical comment and their value in contemporary forensic practice challenged. The question then arises as to what to do with hairs when they are present because this is the reality in many types of crime scenes and scenarios? Are we to simply ignore hairs as an evidence type?

Our aim in this book is to present the case for the continued relevance of hair examination as a valuable biological source that can contribute to assisting to answer questions of identity and questions of what happened or criminalistics. Contemporary forensic practice requires that all potential evidence is evaluated and managed in the context of each unique case. We present a four-level approach to the case management of recovered hairs that can be incorporated into contemporary forensic practice. Our approach focuses on the efficient and accurate selection of hairs for nuclear and mitochondrial DNA analysis whilst also giving due consideration to the criminalistic potential of hairs. We stress the need for thorough and systematic recording of hairs and their microscopic features and on the need to focus on differences to effectively triage recovered hairs. We also discuss how to interpret and report on hair findings to impart to investigators and to the broader legal system what weight should properly be attached to hair findings.

As this is not intended to be a deep academic book but rather a practical book aimed at practitioners, we include a chapter on training of future hair examiners. As hairs are a very visual material, we have included as many images as practical.

One of us (James Robertson [JR]) wrote in 1999 in the *Forensic Examination of Hairs* that there is undoubtably the need for more quality assurance measures, including proficiency testing to ensure the scientific reliability of hair examination. Some 20 years later this remains the case. However, we hope that this book will at least go some way to establishing the ongoing value and validity of hair examinations.

We also wish to thank our colleagues Adine Boehme, for contributing an invited chapter on acquired characteristics, and Melissa Airlie for her useful feedback on various versions of the chapters. Thanks are also due to Dino Todorovic for the original artwork. Unless otherwise specified all images were produced by Elizabeth Brooks. We gratefully acknowledge the donation of hairs from colleagues, family and friends which were used to create the hair images throughout this book and Rachel Read for her hairdressing skills.

Both of us would like to acknowledge our families who have to live hairs with us!!

About the Authors

James Robertson graduated from the University of Glasgow in 1972 with a BSc (Hons) in Agricultural Botany and in 1976 with a PhD in plant physiology. Following a short period as a postdoctoral researcher in London he entered the world of forensic science as a lecturer at the University of Strathclyde where he taught in the Masters of Forensic Science from 1976 to 1985. James worked on a Royal Commission in South Australia in 1983 and in 1985 migrated to Australia to work as a senior forensic scientist in Adelaide before joining the Australian Federal Police (AFP) in 1989 as the first Director of forensic science. During the next 20 years he established the AFP forensic group as a world-renowned forensic organisation.

Despite occupying a senior managerial role, James always maintained his case work competency as a fibre and hair examiner and his interest in all things academic! Hence, it was no surprise on his retirement from the AFP that he returned to an academic role as a Professorial Fellow at the University of Canberra (UC) and Director of the National Centre for Forensic Studies (NCFS). James "retired" (again) in 2019 but remains as a Professor Emeritus at UC.

James has authored and co-authored close to 200 academic papers and edited and co-edited several books on forensic science, including books on fibres and hairs. He is Editor Emeritus of the Australian Journal of Forensic Sciences and a Life Member of the Australian Academy of Forensic Sciences (AAFS) and the Australian and New Zealand Forensic Science Society (ANZFSS) having served both organisations as either the president or the vice-president. James has chaired all of the major forensic advisory groups in Australia during his career. His contributions to forensic science have been formally recognised with the Public Service Medal (PSM), a Member of the Order of Australia (AM) and Doctor of the University of Canberra.

James is a Fellow of the Royal Society of new South Wales (FRSN).

James continues his active interest in, and passion for, the forensic aspects of trace evidence, criminalistics and especially fibres and hairs.

Elizabeth Brooks graduated with a Bachelor of Arts/Biogeography, Australian National University, Australia, in 1978. Before commencing with the Australian Federal Police (AFP) in October 2000, Elizabeth worked for 22 years at the Commonwealth Scientific and Industrial Organization, Division of Entomology providing microscopy/immunohistochemistry expertise in Electron, Light and Confocal Microscopy for the Biotechnology Program specifically and the Division of Entomology generally. This included basic Transmission Electron Microscopy, histology, Scanning Electron Microscopy, immunohistochemical labelling and interpretation, complex time course experiments involving bacteriology, virology, neuropeptides mapping and esterase localisation. During this time, Elizabeth published articles in scientific journals and requested book chapters, relating to the specialised work with virus/insect interactions and pathology. In 2000, Elizabeth joined the AFP as part of the biological criminalistics team undertaking DNA analysis. With a strong background in microscopy, Elizabeth's interest was directed towards forensic hair examination. Training undertaken with Dr James Robertson, National Manager of AFP Forensics and a Master of Applied Science from the University of Canberra, Australia, was completed in 2007—*An appraisal of the use of numerical features in the forensic examination of hair.* Further publications, but with a forensic focus, including another requested book chapter were completed during the years worked at the AFP. Hair examination had become the focus and a way of life! In retirement, Elizabeth maintains her interest and expertise in hair examination by providing hair training courses and technically reviewing cases for Australian jurisdictions, as requested.

CONTRIBUTING AUTHOR BIOGRAPHY

Adine Boehme graduated with a Bachelor of Forensic Studies, University of Canberra, Australia, in 2006 and followed with a Bachelor of Applied Science (Hons) in 2007. This honours project was a forensic hair related one that resulted in a publication analysing the transfer and persistence of animal hair in a forensic context. In June 2007, Adine joined the Australian Federal Police (AFP) as part of the biological criminalistics team, working under the guidance of Elizabeth Brooks. With an established interest in forensic hair examination from her honours project, Adine decided to undertake the necessary training to make forensic hair examination her specialisation. Training over several years with Elizabeth Brooks and Dr James Robertson, Adine achieved her expert status with considerable success. Adine is now the one of the only two scientists practising Forensic Hair Examination for the AFP. As there are only three qualified hair examiners currently employed in Australia, Adine and her colleague at the AFP extend hair examination services to the AFP, Interstate policing and International policing clients.

CHAPTER 1

Historical Context to Contemporary Practice in Hair Examinations

1.1 INTRODUCTION

The purpose of this chapter is to provide the reader with relevant background about the examination of hairs in a forensic context. It is not our intention to comprehensively review the detailed history of how the examination of hairs has developed over the last hundred or so years since the first formal book on the examination of hairs appeared in 1910 (Lambert and Balthazard, 1910). Our coverage will focus on the background relevant to understanding why some observers and commentators have expressed largely negative opinions on the value of microscopic examination of hairs and to put the case forward as to why the examination of hairs still has value in contemporary forensic practice.

For example, in evidence to an enquiry into the wrongful conviction of Driskell for murder (the Driskell Inquiry) in which hair evidence played a critical role, Peter Neufeld stated that "it is speculative, actually, to suggest that hair microscopy can play a useful role in the forensic science or criminal adjudicatory process". He conceded that hair examination "could be useful as a screen in cases of obvious visual exclusion" but opined that "because hair has no database from which one can give a statistical opinion" he could see no reason why it "continues to serve a significant role" (Anon, 2006). In the same enquiry, Tilstone concluded that "the only thing I think is scientifically justifiable as a conclusion from a forensic lab in regard to microscopical hair examination is to say the questioned hairs could not have come from the known source". However, he did add that "in the absence of being able to make that definitive assertion, a report then has to go ahead and explain what the limitations are on any implications or inferences that someone would care to draw in regard to the possibility of them having a common origin" (Anon, 2006).

DOI: 10.4324/9781315210650-1

Lucas gave a slightly more positive assessment of the potential value of hair examination in a report prepared for the Driskell Inquiry concluding that "microscopic hair comparison continues to be a useful technique in forensic science for exclusionary purposes and may be helpful for inclusionary purposes in certain circumstances" (Lucas, 2007). The Driskell case has special significance as the hair examination was conducted by an examiner working for the forensic laboratories of the Royal Canadian Mounted Police (RCMP), the laboratory system for whom the late Barry Gaudette worked and produced a series of papers in the 1970s and 1980s on the evidential value of hair examination (Gaudette, 1999). The significance of this work will become clear later in this chapter.

The much quoted 2009 report "Strengthening Forensic Science in the United States: A Path Forward" by the US National Research Council (hereinafter the NAS Report) also commented on hair examination, stating that forensic hair examiners generally recognise that various physical characteristics of hairs can be identified and are sufficiently different among individuals, and that they can be useful in including, or excluding, certain persons from the pool of possible sources of the hair. The results of analyses from hair comparisons are typically accepted as class associations; that is, a conclusion of a "match" means only that the hair could have come from any person whose hair exhibited—within some levels of measurement uncertainties—the same microscopic characteristics, but it cannot uniquely identify one person. However, this information might be sufficiently useful to "narrow the pool" by excluding certain persons as sources of the hair (Anon, 2009, p. 156).

The NAS report (pp. 160–161) also identified a number of concerns about microscopic hair analysis, specifically,

- no scientifically accepted statistics exist for the frequency with which characteristics of hair are distributed in the population,
- there appear to be no uniform standards on the number of features on which hairs must agree before an examiner may declare a "'match",
- the categorisation of hair features depends heavily on examiner proficiency and practical experience,
- imprecise reporting terminology can lead to misunderstanding and wrongly imply individualisation, and
- although microscopy and mitochondrial-DNA (mt-DNA) analysis can be used in tandem and may add to one another's value for classifying a common source ... no studies have been performed specifically to quantify the reliability of their joint use.

The 2016 report to the President (of the United States of America) by the Presidents' Council of Advisors on Science and Technology (PCAST) considered the scientific validity of feature-comparison methods, including hair analysis (Anon, 2016a). In assessing scientific standards for scientific validity, PCAST distinguished between two types of scientific validity terming these as **foundational validity** and **validity as applied**. They defined these as follows:

- *Foundational validity* for a forensic science method requires that it be shown, based on empirical studies, to be *repeatable, reproducible*, and *accurate*, at levels that have been measured and *are appropriate to the intended application* (our emphasis). Foundational validity, then, means that a method can, *in principle*, be reliable.
- *Validity as applied* means that the method has been reliably applied *in practice*.

The authors of the PCAST report specifically drew attention to the need for particular scrutiny for subjective methods inherent in forensic feature comparisons because "their heavy reliance on human judgment means they are especially vulnerable to human error, inconsistency across examiners, and cognitive bias. In the forensic feature-comparison disciplines, cognitive bias includes the phenomena that, in certain settings, humans may tend naturally to focus on similarities between samples and discount differences and may also be influenced by extraneous information and external pressures about a case".

The report further asserts that "without appropriate estimates of accuracy, an examiner's statement that two samples are similar—or even indistinguishable—is scientifically meaningless: it has no probative value, and considerable potential for prejudicial impact".

The report also rejects experience, judgement and good professional practice having any role as a substitute for empirical evidence supporting foundational validity and reliability.

The report goes on to analyse how seven feature-comparison methods measure up in terms of their criteria of foundational and applied validity.

With respect to hair examination, PCAST concedes that it did not undertake a comprehensive review, instead choosing to restrict consideration of hairs to a review of documentation released by the US Department of Justice (DoJ) supporting proposed uniform language guidelines for testimony and reports (Anon, 2016b). Although the primary purpose of the DoJ supporting documentation was not intended to address the validity and/or reliability of the hair examination discipline, the supporting documentation did state that "microscopic hair comparison has been demonstrated to be a valid and reliable scientific methodology", whilst noting that "microscopic hair comparisons alone cannot lead to personal identification and it is crucial that this limitation be conveyed both in the written report and in testimony". Whilst recognising the constraints on the DoJ, the PCAST report proceeds to systematically dismantle the "evidence" cited in the DoJ document to support the scientific validity for hair examinations. The DoJ analysis centred primarily on studies by Gaudette in the 1970s (Gaudette, 1999) and on a 2002 Federal Bureau of Investigation (FBI) study in which mt-DNA analysis showed that 11% of hairs, previously found to be indistinguishable by microscopic examination, actually came from different individuals (Houck and Budowle, 2002).

Clearly the authors of the PCAST report do not believe hair examination meets the first barrier of foundational validity.

Furthermore, serious concerns have been raised over hair examination within the FBI before the introduction of mt-DNA testing in 2000. These concerns centre on the testimony given by FBI examiners as to the strength of hair evidence to individualise.

The enquiry into FBI testimony on microscopic hair analysis pre-2000 raises deep concerns as to the "validity as applied" criteria, at least as applied in the FBI laboratory in the relevant period under review (Anon, 2018a).

Hence, it must be asked whether there is any ongoing value to be attached to the examination of hairs in a forensic context, or should forensic organisations cut their losses and, as many have already done, exit the field?

We will present the case why hair examination should not only be retained in current forensic practice, but also what is necessary to address the "foundational validity" criteria and ensure it is properly applied to meet the "validity as applied" criteria.

1.2 A BRIEF HISTORY OF THE LAST
100 YEARS OF HAIR EXAMINATION

To appreciate the current status of hair examination, and what the future may hold, it is important to think more broadly about how science, and bioscience, has evolved in the last 100 years; after all, hairs are just one form of biological material. In our view, this period of history can be broken up into *three* broad *eras* with some overlaps. We are now in an overlap period with the next or the emerging fourth era!

Era 1 started well before the early 1900s and lasted until the 1950s. From a technology perspective it was the **era of observation and microscopy** as there were few of the analytical technologies taken for granted in today's laboratories. Putting to one side for the moment any consideration of the *value* of microscopically observable features, there is no doubt that hairs possess a range of interesting characteristics that can be seen through appropriate microscopic observation. Classic textbooks such as Lambert and Balthazard in 1910 include very detailed hand-drawn figures showing many of the features relied upon today

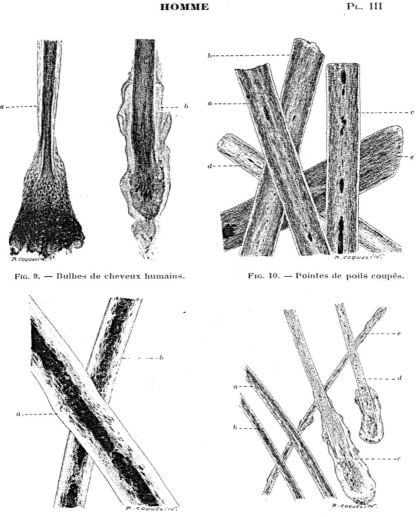

HOMME   PL. III

FIG. 9. — Bulbes de cheveux humains.   FIG. 10. — Pointes de poils coupés.

FIG. 11. — Poils de moustache d'homme   FIG. 12. — Poils de fœtus humain.
(pigmentophages).

FIGURE 1.1 Plate III of human hair drawings from Lambert and Balthazard (1910). The microscopic features of hairs seen today were observable in 1910.

(see Figure 1.1). The latter text, along with many studies published in the 1920s and 1930s, with the names of Hausman and Trotter featuring prominently (Robertson, 1999a), is the foundation for our understanding of the microscopic features found in hairs. Neither Hausman nor Trotter were forensic scientists and their studies were published in journals such as the *American Journal of Physical Anthropology* or the *American Naturalist*. In the same era, the eminent Scottish medical practitioner, John Glaister, also published on the use of hair examinations in forensic medicine (Glaister, 1931). His Scottish colleague, and fellow forensic pathologist, Sydney Smith, in his autobiography, details the case of the murder of an eight-year-old girl, Helen Priestly, which included hair examinations and comparison. He concluded that "as far as I could judge by examination in a comparison microscope, the hairs from the sack and those from Mrs Donald (the accused) were identical in every detail but this could not be regarded as proof that it was Mrs Donald's hair". He further added that "unless there is something very characteristic about a certain head of hair, not much reliance can be placed upon it for the purposes of identification" (Smith, 1982). Perhaps some forensic examiners from later generations would have benefitted from paying more attention to the work of Glaister and Smith?

Not much changed in the world of hair examination in the next 30 years, except for the emergence and development of the role of the "forensic scientist" who, to a large extent, took over laboratory examinations of hairs from forensic pathologists.

Era 2 started in the 1970s where the value of hair examinations was certainly under the microscope with the late Stuart Kind questioning the information content gained from the microscopical comparison of hair samples (Kind and Owens, 1977). Two key events were to drive a period of intense interest in the forensic examination of hairs. The first event was the publication of the first paper (of many!) by Barry Gaudette titled "An attempt at determining probabilities in human scalp hair comparison" (Gaudette and Keeping, 1974). Gaudette's finding that in head hair studies, a total of 370,230 inter-comparisons were conducted, with only nine pairs of hairs that could not be distinguished corresponded to a false positive rate of about 1 in 40,500 and became the catalyst to promote a vigorous re-assessment of the value of microscopic examination of hairs. The PCAST report, noting that this and subsequent studies by Gaudette had been the subject of strong criticism, rejected this work as giving a sound scientific basis for concluding that microscopic hair examination was a valid and reliable process (Anon, 2016a).

Gaudette (1999) and Wickenheiser and Hepworth (1990) had attempted to address the critical commentary of the Gaudette studies, principally from Barnett and Ogle (1982). Gaudette (1999) accepted that the so-called "average probabilities" had their limitations but believed that they "are likely all that will ever be available with forensic hair comparison". Critically, he pointed out that *even* these average probabilities depended on examiner error being very low and that this would only be met when a well-trained, qualified examiner carefully conducted the examination. Furthermore, Gaudette never suggested that these numbers be used in isolation of a range of other factors that needed to be considered that would tend to weaken or strengthen hair conclusions (Gaudette, 1999). We discuss these factors in Chapter 6 of this book. He even discussed how the approach he advocated could be adapted to a Bayesian approach to evidential value that was emerging in this same era.

The second event of this era was technology based and involved developments in analytical chemistry that swept through laboratory science in the 1960s and 1970s. With respect to the biosciences, the critical technology was electrophoresis (Righetti, 2005; Vesterberg, 1989). Originating from the work of Tiselius in the 1930s (Tiselius, 1937), it was the introduction of starch gels, and later polyacrylamide gels, which enabled the efficient separation of proteins. The development of isoelectric focusing (IEF) in the 1960s was also important for forensic applications. Murch and Budowle (1986) reviewed the applications for IEF in forensic serology and demonstrated the central role for this technique *even* as we were on the cusp of the next seismic change in forensic biology, the ability to analyse DNA from forensically relevant samples. Of course, electrophoresis remains the core technology to separate DNA until today.

With respect to hairs, electrophoresis allowed forensic scientists to look at multiple enzyme typing of sheath cells associated with the roots of human hairs (Lawton and Sutton, 1982). Although this was a real breakthrough, its utility was restricted as sheath cells are normally only present in some growing, or anagen, hairs, usually indicating forcible removal. Arguably, the aspect of hairs that attracted greater interest and excitement in the late 1970s and early 1980s was the use of electrophoresis to examine matrix proteins as a potential tool for identification (Lee *et al*, 1978; Marshall *et al*, 1985). These groups of researchers used two-dimensional IEF to produce complex patterns of matrix proteins. Although the "potential" of this approach was never realised, in part simply because DNA technology emerged very soon thereafter, "what goes around, comes around" and a recent 2016 paper raises again the potential for protein-based human identification using hair shaft proteome (Parker *et al*, 2016).

One of us (James Robertson [JR]) first published on forensic hair examination in the early 1980s (Robertson, 1982) and then collaborated with Aitken to look at the value of microscopic features in the examination of human head hairs (Aitken and Robertson, 1986), developing an analytical scheme based on an assessment of contemporary practice (Robertson and Aitken, 1986). We also contributed to the debate on the use of probabilities and human hair comparisons (Aitken and Robertson, 1987).

Robertson (1982) concluded that the reasons why it had proven so difficult to improve the discriminating power of hair examinations included that

- the microscopic features of hair were difficult to assess in an objective way,
- there was considerable variation in features within a sample of hairs from one individual and that intra-variation could be as great as inter-individual variation, and
- there was little population data relating to the frequency of occurrence of different features in hairs.

Robertson further commented on the term's "objective" and "subjective" in relation to microscopic features arguing that it would be more useful to accept that microscopic features are not objective in a numerical sense but that they could be assessed objectively in terms of

- the ability of one person to reach the same decision, given the same feature or hairs to examine on several occasions, and
- the ability of different analysts to reach the same decision.

However, Robertson argued that this would require clearer and more adequate definitions of the features to be assessed.

Hence, at the start of the 1980s, there was considerable debate as to the value of hair examination, fuelled in large measure by the seminal 1974 paper by Gaudette (Gaudette and Keeping, 1974) and a real sense that perhaps protein analysis could be the way forward to achieving a more objective assessment of individuality. There followed several major conferences and meetings that brought together forensic scientists across the world to discuss hair examination. It was, put simply, an exciting time to be involved with hair examination. These meetings included a 1983 FBI-sponsored workshop that established a committee for hair examination—the forerunner of the Scientific Working Group for Materials Analysis—Hair Subcommittee (SWGMAT) (Anon, 1985a).

In 1984, along with Manfred Wittig from Germany, one of us (JR) organised a full-day section on hairs at the Oxford meeting of the International Association of Forensic Sciences (IAFS), and, in 1985, a further meeting, hosted by the FBI, brought together 172 scientists from around the world that resulted in a 221-page proceeding (Anon, 1985b). At this meeting, subcommittees, formed at the 1983 meeting, reported on report writing, conclusions and court testimony and on draft guidelines for the establishment of quality assurance programmes for the forensic comparison of human hair. The future seemed only to hold promise! So why is it that some 30 years later, far from this promise being realised, to many observers, hair examination is in terminal decline? To answer this question, it is required, at least in part, dealing with the **3rd Era, the emergence of DNA analysis!**

It is well beyond the scope of this chapter to review the evolution of DNA analysis since its first "forensic" application (Gill and Werrett, 1987). When DNA analysis is discussed, it is often referred to as the gold standard for forensic science. Of course, the compelling attraction of DNA analysis for hair examination is that it potentially addresses the "identification issue". Interestingly, the first forensic application for DNA was an exclusion, not an inclusion! One of the co-authors of that first paper was of course, Peter Gill. Since then Peter has been a prolific author of papers dealing with technical and interpretation aspects of DNA. His 2014 book on misleading evidence (Gill, 2014) deals with the dangers of an uncritical application of DNA analysis that does not address the "broader issues". Gill points out that "the fact that a DNA profile may match a defendant is of secondary interest to the questions: 'how' and 'when' did the DNA transfer take place?" He goes on to comment that "it is not the 'fact' of a matching DNA profile that is of primary interest, rather it is the 'context' or the 'relevance of the DNA profile to the crime event itself". Gill was largely the architect of the so-called low-copy number procedures (Gill, 2001), but he cautions of the dangers in "overstepping boundaries of knowledge" and the need to understand transfer and persistence of trace biological materials, and especially skin cells. Gill interestingly points out that the power of DNA (databases) is "the ability to *eliminate* (our emphasis) large numbers of innocent people from an investigation".

Our point is that even with DNA there has been *too* strong a focus on identification and an almost blind acceptance that if the DNA number is large enough, then this equates to guilt. One only needs to read the Vincent Inquiry report on the wrongful conviction of Jama in Australia to fully appreciate this point (Vincent, 2010).

Hence, the effective use of DNA, or to be more accurate, biological material and often trace biological material, requires an understanding of **criminalistics** as well as the foundational validity of the DNA process. Validity as applied is not limited to just meeting the required technical standards. In the context of criminalistics, hairs are an excellent material as they are visible, relatively easy to collect and their location may yield useful information with respect to Gill's "how" and "when" questions. For a more in-depth consideration of the criminalistics of trace the reader is referred to Roux *et al* (2016).

With respect to hairs, the physiology of the life cycle of hairs from inception to eventual loss is well understood (Harding and Rogers, 1999). Microscopic examination of hair recovered in the context of an investigation will generally allow an interpretation of "how" a hair was lost based on its inherent characteristics (growth phase as determined by root appearance) and acquired characteristics (such as wear or damage). Research on the transfer and persistence of hairs (Dachs *et al*, 2003) assist the forensic criminalist interpret the "when" question. In brief (see Chapter 2 for a more detailed treatment), whilst the life cycle of a hair is a continuous process, human hairs are classified based on the appearance of their roots as growing (anagen) or dead (telogen) with an intermediate short-lived stage called catagen. Not all recovered hairs will have a root present—they may have been broken. However, when a root is present, this should be assessed prior to decisions as to whether to conduct further microscopic examination. If the root is either in the catagen or anagen growth phase, then it is suitable for routine nuclear (nu) DNA testing. In our view, even in these circumstances, it is important to record basic information about the hair at the level of a low-power microscopic examination (LPM). Provided a relatively fresh root is present, close to a 100% success rate should be achieved for nu-DNA testing of non-telogen hairs. In these circumstances' hairs are an excellent biological material as they can provide information across the three levels of "activity", "source" and "sub-source" within the "framework of propositions" proposed by Cook *et al* (1998).

The limitation is that whilst over 90% of human head hairs are in anagen or catagen growth phase, these hairs are tightly held in the dermis, and, hence are *not* the hairs normally recovered for forensic examination. By contrast, hairs in the telogen growth phase are loosely held closer to the skin surface, are more easily removed by grooming or other means and, hence, make up about 95% of recovered hairs (Robertson, 1999b). Telogen hair roots, and the hair shaft of all human hairs, do have nu-DNA present *but* in a highly dehydrated environment that makes its extraction a major challenge. Even if this nu-DNA can be extracted, it is likely to be highly degraded and unsuitable for routine short tandem repeat (STR) based DNA analysis (Edson *et al*, 2013). However, this does not mean that a "reportable" nu-DNA result cannot be obtained from some telogen hairs. The key to success appears to lie in the presence of detectable nuclei, in the main associated with cellular debris attached to the hair root (Brooks *et al*, 2010; Haines and Linacre, 2016). Success still requires using a low template DNA approach to analyse the DNA (McNevin *et al*, 2015). The use of a direct amplification approach is reported to improve the percentage success rate for telogen hairs to around 30% (Ottens *et al*, 2013). The nu-DNA may be exogenous, associated with the cellular debris, and not endogenous, belonging to the actual hair and, hence, extracted nu-DNA may be from a mix of exogenous and endogenous sources. In this sense, the hair may be considered as merely a template for secondary biological trace. The question of whether extractable nu-DNA from hair shafts is exogenous or endogenous remains uncertain, although McNevin *et al* (2005) believe that it is most likely exogenous. The same authors have also shown that successful DNA typing of telogen hairs is donor dependent (McNevin *et al*, 2005).

To summarise the current status of nu-DNA analysis for hairs, hairs in the anagen growth phase are an excellent source of nu-DNA and should result in close to 100% success rates. For hairs in the telogen growth phase, nu-DNA analysis is *not* a routine procedure. Telogen hairs need to be treated as a specialist DNA technique, and this may involve pre-assessment for the presence of nuclei and/or modified extraction or even direct amplification. Current success rates for a reportable nu-DNA profile could be expected to be between 10% and 30% of recovered telogen hairs. Obtaining a reportable nu-DNA profile from telogen hairs is likely to remain challenging for the foreseeable future. However, the use of more sensitive commercial DNA kits and/or application of massively parallel sequencing (MPS) may further increase success rates.

All recovered hairs should be treated as individual hairs. However, in many cases several hairs are recovered. Microscopic examination may provide reasonable grounds to treat multiple recovered hairs as having a possible common origin. In these cases, it only needs one hair to yield a reportable nu-DNA profile to provide identification level information.

Thus far only nuclear DNA has been considered. However, hairs are also an excellent biological material for mt-DNA analysis (Melton *et al*, 2005). Mt-DNA analysis is essentially destructive; however, only a short length of hair shaft is required for analysis. Melton *et al* (2012a) showed that 19 of 21 hairs, less than 2 mm in length, gave at least a partial profile when analysing hairs from a 19-year-old homicide. As is well known, and appreciated, mt-DNA has several limitations in the information it provides compared to nu-DNA testing. These limitations include issues of interpretation, specifically heteroplasmy in hairs and the fact mt-DNA is inherited only through the maternal line. Mt-DNA results require specialist interpretation and no DNA databases *equivalent* to those used for nu-DNA exist (Melton *et al*, 2012b). Mt-DNA analysis is also more time consuming and more expensive than nu-DNA testing, much of which is now conducted with automated laboratory protocols. Melton *et al* (2012b) conclude that "the most significant challenge for mt-DNA analysis remains the high cost and low throughput for evidentiary samples". Hence, it is important that hairs are not submitted for mt-DNA analysis unless they have been examined with some level of microscopic examination. Melton *et al* (2012b) also point to the emergence of MPS as holding out promise in improving interpretation issues around mixtures and heteroplasmy. This will involve the use of computer algorithms to deconvolute data in what is the emerging **4th Era** of **"big data" and "bioinformatics"** (Marx, 2013).

Finally, with respect to DNA analysis, the PCAST report noted that a "particularly concerning aspect of the DoJ supporting documentation is its treatment of the FBI study on hair examination" (Anon, 2016a, p. 121). The writers of the PCAST report concluded that this study found that in 9 of 80 cases (11%), in which hairs had been found to be microscopically indistinguishable, the DNA analysis showed that the hairs came from different individuals. The writers of the PCAST report were "surprised" and suggested the DoJ report completely ignored this finding. The FBI study was in fact a paper by Houck and Budowle (2002). It is the view of these authors that "these nine mt-DNA exclusions should not be construed as a false positive rate for the microscopic method or a false exclusion rate for mt-DNA typing". This comes down to one issue, and that is, how do you define an error or mistake? In the context of the microscopic examination of hairs, it raises issues of type 1 and type 2 errors and the purpose for microscopic examination that go to the very heart of this review. We shall return to this core issue in the final section of this chapter. For the present, suffice to say, that the major limitation of the PCAST report is the limited and singular focus on the only role of hair examination being "individualisation".

However, in terms of evaluating the foundational validity of hairs, if hairs are viewed *only* as a biological source of DNA, then if DNA science is "the gold standard" so then are hairs.

1.3 THE CRIMINALISTICS VALUE OF HAIRS IN THE CONTEXT OF CONTEMPORARY CASE MANAGEMENT

For some, hairs are an "inconvenient truth". However inconvenient they may be, hairs are, if not ubiquitously present, an extremely common occurrence in forensic context. The reason for this is simple; animals and humans lose or shed their hairs. For humans, in the order of 100–150 scalp hairs are lost each day, not to mention hairs from other parts of the body (Robertson, 1999b). As a physical material, hairs may be found in almost all, if not all, types of criminal investigations. A key skill for the crime scene examiner or forensic examiner at the scene, and in the initial examinations in the laboratory, is to *recognise* what may be present, but, equally, what may eventually provide useful information at the investigative and criminal justice process stages. If the view is held that hairs provide a poor return in terms of the investment of time, relative to the value of the potential outcomes, perhaps it would not be surprising if little regard or effort was made to record or recover hairs. Hence, as for any type of physical material, it is important that the "collectors" of such materials have a well-informed and balanced approach to "evidence" recovery. One thing can be stated with certainty: if hairs are not collected, they cannot be examined. In recovering hairs, consideration should be given to their location and how they may be held or situated, as this may provide important criminalistics insights and contribute to event reconstruction. Robertson (1999b) describes several case examples in which the location of hairs was as important as identifying the possible donor. There will also be many cases where the need to identify a source is limited to a closed population or a small number of individuals, such as, who was the driver in a vehicle accident.

Today, there is an understandable focus on biometrics, more broadly and specifically, early identification of suspects or victims to assist the investigation phase. Often criticised in the past for taking too long to provide information, contemporary forensic practices are focussed on providing this type of information as quickly as possible. As has been demonstrated in the last section, hairs can be an excellent biological material that can provide DNA results contributing to "identification", and hence, simply on this basis alone hairs should not be ignored. They have additional benefits of being less likely to produce mixed DNA profiles and providing a broader criminalistics contribution. However, as previously covered, not all hairs will provide a nu-DNA result. The question then is, to what extent should hairs be further examined? To answer this question, it is critical to be clear about the purpose of further examination. Some organisations will decide that only hairs that are obviously suitable for routine nu-DNA testing will be processed. In our view, this is preferable to a half-hearted commitment to further examinations. This can only continue to do damage to the credibility of hair examinations.

If hairs have been recovered at the crime scene, or during a subsequent laboratory-based search and recovery procedure, then a decision will need to be made as to whether these hairs should be subjected to further examination. This should be part of the case management protocol for an organisation. Many organisations now operate one form or another of a triage approach in the early stages of an investigation aimed at improving both the efficiency of processes and delivering timely outcomes. The management challenge remains ensuring that *effective* outcomes are not confused solely with *early* outcomes—some detailed tests simply take time. Decision making in these early stages is not trivial and poor decision-making can have very serious downline implications. However, assuming good practice and decision-making, it would only be in rare circumstances that all hairs will be examined in detail.

As well as being a potential source of nu-DNA, acquired characteristics may provide useful information. For example, was a hair forcibly removed or does damage to the hair shaft indicate the nature of the damage? Following visual examination, hairs may be selected for more detailed examination using microscopy. The first stage in microscopic examination should involve LPM, usually with a stereo microscope with a zoom total magnification of up to 40× or 50×. Hairs are best examined at the LPM level using a ring light source to provide even epi-illumination. It is also preferable if hairs are placed in a suitable mountant such as Histomount™. At this level of magnification, details such as root growth phase, tip appearance, overall colour, "acquired" colour and the presence of damage or parasites are noted. Robertson (1999b) describes this process emphasising the value of a **checklist** to ensure thorough and detailed recording of information. This attention to detail carries on to hairs examined at high-power magnification using transmitted light microscopy (TLM).

A remarkable feature of the hair evidence in the previously mentioned Driskell case (Anon, 2006) was that not only did the examiner not make detailed notes of microscopic features but that the RCMP Hair and Fibre Section Methods Manual directed hair examiners *not* to make detailed notes with respect to questioned hairs. In fairness to the RCMP, this was not an unusual situation even as late as the early 1990s. When Robertson and Aitken (1986) surveyed attitudes to the use of checklists and recording of details of hair examinations, several respondents indicated they did not record detailed information, with some offering as a reason, that the defence could misuse their examinations!

If there is no known (sometimes called exemplar) hair sample from a victim or suspect, then this may end the forensic examination. In most cases, there will be a known sample from a person(s) of interest, and provided this known sample is representative and adequate (sadly *not* always the case), hairs from the known sample can be selected and each hair individually examined. With one or more known samples, the forensic examiner will then consider what features are different between samples that may form the basis of differentiation, note, *not* individualisation. Similarly, the features of recovered hairs can be assessed against the known hairs and known samples excluded, or not, as possible sources of recovered hairs.

When humans look at each other, one of the most obvious features is hair colour. In physical anthropology, the Fischer-Saller scale is used to assess visual hair colour (Fischer and Saller, 1928). This scale classifies hairs into eight groups, these being, very light blond, light blond, blond, dark blond, brown, dark brown/black, red and red blonde. If we substitute "yellow" as an actual colour for the visual term "blond" and remove shade (light or dark), then what is left is yellow, brown, red and black. This is unsurprising as hair colour derives from relative amounts of only two pigments, eumelanin (producing brown and black colours) and phaeomelanin (producing yellow and red colours). The physiological processes through which hair pigments originate in the hair follicle are well understood. Melanocytes produce melanosomes, which in turn syn-thesise pigment granules (Tobin *et al*, 2005). The underlying molecular genetics of human pigment diversity have also been documented (Sturm, 2009). Hence, there is a sound scientific "foundation" that exists out-side of forensic application, underpinning how hair colour is formed, and even how it changes due to aging (Tobin, 2009). Robertson (1999b) discusses how various authors have attempted to classify hair colour for forensic application. Many of the schemes proposed the use of numer-ous colours. Robertson settled on only five groups, these being colour-less, yellow, brown, reddish and black. Further potential to discriminate would consider shade with light, medium or dark. What this scheme attempts to do is minimise variation in classification by an examiner or examiners to achieve greater consistency between examiners and reduce subjectivity. Of course, the above "colours" are as assessed with LPM. In the scheme used by Robertson (1999b), six of the fourteen groups of fea-tures assessed with TLM attempt to further describe the visual appear-ance of pigmentation as seen in the shaft of human hairs. These six groups are pigment density, distribution, and shape and size of pigment granules and aggregates. The remaining eight groups of features assessed include discrete features such as the presence or absence of ovoid bod-ies and cortical fusi, descriptions of medulla, presence or absence, and where present, distribution and appearance, and features associated with the cortex and cuticle.

With respect to pigment features, the challenge in assessing these is achieving consistency between examiners. Even with a single examiner, the primary purpose in adopting a checklist approach remains ensuring a systematic and detailed examination process. As detailed examinations are without question time consuming, they will usually only be conducted on hairs after a screening or triage approach with LPM. Hence, hairs that are excluded at that stage would not normally be examined with TLM. During TLM examinations, the examiner attempts to describe a holistic visual image seen in three dimensions and, hence, never all in focus at one time. A checklist can never be more than an attempt to cap-ture "information" in a one-dimensional format. To capture this three-dimensional information in one dimension (or field of focus), Brooks *et al* (2011) applied an auto montage approach that produced a "stacked" composite image. Their work went on to use these composite images to explore the potential for using digital image analysis to extract pigment pattern images for pattern matching. Although this approach has not yet progressed further, it was the first (and to our knowledge, only) attempt to develop an operator-independent method of objectively capturing pig-ment patterns in human hairs.

This goes to the heart of criticisms raised in the PCAST report and over a long period by many other commentators, that is, reliability based on reproducible performance. Dror (2016) comments on the lack of reliability between forensic experts as "concerning" but says that the lack of reliability within experts is "alarming". In relation to expert decision-making, he observes that this should be driven by "data and **expertise**" (our emphasis!) and that, "the same expert, looking at an identical situation, should generally reach the same conclusion". As we have stressed several times already, the objective of the microscopic examination process should be systematic and thorough and should capture as comprehensive a picture as possible of the complex information package that constitutes a hair. In forensic processes where there are known hair samples, against which recovered hairs can be compared, by working out where the differences lie between known hairs, recovered hairs can be eliminated based on understanding what constitutes a meaningful difference. It is extremely disappointing that the proposed uniform language for testimony and reports from the United States retains the use of terms such as "similarities" and "microscopically consistent with" (Anon, 2016b). Similar implies no level of discrimination. For example, two almost featureless, colourless hairs will be similar, but the reality is that the inherent "information" is also very low. On the other hand, when two well-featured hairs are examined and compared, the "information package" upon which an exclusion is possible is much more meaningful.

In discussing his "hierarchy of expert performance", Dror (2016) draws attention not only to the issue of "reliability" but also, to what he terms, "biasability". Without entering an in-depth discussion about bias, we draw attention to one aspect raised in this paper where he states that when "conclusions drive observations (rather than vice versa), confirmation bias kicks in", arguing that conceptually and practically, observations should come first. He goes on to further state that "observations must start with the evidence from the crime scene and then, thereafter, observations from the suspect".

In principle, we agree that it is desirable to examine recovered hairs first and to have recorded their microscopic features at the LPM level. However, it would be impractical and inefficient to record the features of all recovered hairs at the TLM level before examining any known hairs from victims or suspects, as the primary purpose of examinations at the LPM level are exclusions. In some cases, known questioned/recovered hairs are collected from numerous locations as an investigation develops. The hair examiner would then examine these hairs to determine whether any known sources can be excluded. This may require looking at hairs at the TLM level. Hence, at some point the examiner will have a detailed knowledge of known samples before a recovered hair is examined at the TLM level.

Provided the examiner records microscopic features in a thorough and detailed manner with a focus on exclusion as their primary goal, we believe this approach is justified and minimises potential bias.

We agree that hair examiners should of course be protected as far as possible from task-irrelevant information. Dror proposes that the application of a linear sequential unmasking (LSU) procedure ensures that the approach moves from the suspect to the evidence (Dror *et al*, 2015). We discuss these issues in greater detail in Chapters 5 and 6 of this book.

In our opinion, as all hairs have the potential to yield a DNA result (nuclear or mitochondrial) the goal of a hair examination protocol should now be to

- at the earliest opportunity select hairs suitable for routine nu-DNA analysis,
- examine remaining hairs at LPM to eliminate recovered hairs based on differences from known hair samples,
- if the organisation has in place protocols for non-routine nu-DNA testing (such as counting nuclei or direct amplification), select hairs that have not been excluded based on LPM for such testing, and
- if the organisation has examiners authorised to conduct TLM, examine known and recovered hairs and compare with a view to eliminating recovered hairs based on there being a meaningful difference.

Because mt-DNA testing is available, the balance has changed from "if in doubt, eliminate" to, within reason, "if in doubt, retain". Under this approach it is *not* an error if a hair is submitted for mt-DNA testing and then eliminated based on the mt-DNA result. Equally, for hairs with a low information value (colourless or conversely, opaque), failure to exclude based on microscopy is also *not* an error.

Finally, the ability to reach a meaningful conclusion will be influenced by the nature of the recovered hair and the quality of the known samples. These reasons were discussed by Houck and Budowle in their 2002 paper which is mistakenly used to suggest a significant error rate for hair microscopy. As these authors quite correctly point out microscopy and mt-DNA analysis are complementary. Houck and Budowle (2002) also argue that microscopy should not be viewed as merely a "screening test" and mt-DNA as a "confirmatory test". As they point out, both methods can provide probative information. As previously discussed, viewed from a criminalistics perspective, hairs as a physical material can provide useful information at an investigative and evidence level, especially, in helping answer the "when" and "what happened" questions.

1.4 IMPLICATIONS ARISING FROM THE REVIEW INTO THE FBI APPROACH TO HAIR REPORTING

Finally, a brief mention of the FBI/DoJ review into microscopic hair comparisons conducted within the FBI from the early 1980s until the widespread introduction of mt-DNA testing in 2000. To date, no detailed report from this review has been published with the only information available appearing in press releases (Anon, 2015a). As of May 2015, only 500 of 3,000 cases, meeting the review criteria, had been reviewed. Of these cases, the relevant press release stated that "examiners testimony in at least 90 percent of trial transcripts … contained erroneous statements". Twenty-six of 28 FBI agents/analysts provided either testimony with erroneous statements or submitted laboratory reports with erroneous statements. It appears unlikely that a final report from this review will be forthcoming. However, the FBI in 2017 commissioned a "root and cultural causes" review to examine the likely reasons that contributed to the report and testimony errors occurring and not being abated by FBI management (Anon, 2018b). As of June 2018, the FBI had found errors in 856 of 1,729 reports and in 450 of 484 transcripts with 31 of 35 examiners having made these errors (Anon, 2018b). What constituted an error was defined by the FBI as follows:

> *Error type 1:* The examiner stated or implied that the evidentiary hair could be associated with a specific individual to the exclusion of all others.
> *Error type 2:* The examiner assigned to the positive association a statistical weight or probability or provided a likelihood that the questioned hair had originated from a particular source, or an opinion as to the likelihood or rareness of the positive association that could lead to jury belief that a valid statistical weigh can be assigned to a microscopic hair association.
> *Error type 3:* The examiner cites the number of cases or hair analyses worked in the laboratory and the number of samples from different individuals that could not be distinguished from one another as a predictive value to bolster the conclusion that a hair belongs to a specific individual.

The root and cause review found fundamental management system weaknesses that included the FBI not having formalised methods or formally establishing criteria for determining errors and monitoring against these criteria. The FBI laboratory also exhibited overconfidence in that they did not need outside expertise and did not see the value in formalised procedures. As a result of these cultural weaknesses, the FBI did not have sufficiently specific guidance needed for reports and testimony during the period reviewed. The review also found a number of other cultural causes that contributed to errors. These included FBI agent-examiners acting like detectives, and not impartial scientists, and a culture that did not value a questioning and learning environment but rather one of *thoughtful-compliance* relating to report and testimony statements.

Note that this review did not examine the underlying science being limited in its brief to only examining reports and testimony.

We have already offered some thoughts on the dynamic that defined hair examination in the early 1980s culminating in the FBI-sponsored meeting in Quantico in May 1983. This was the first time that one of us (JR) had met many of the FBI hair examiners. Almost exclusively, these examiners were "agents". All had degrees but not necessarily in science. For some agents this was their long-term career but, for many, they would spend a limited period as an examiner before returning to a field agent role. When discussing hair examination with many of these individuals, one could only be struck by the almost cult-like responses and belief in what they were doing. When asked about publishing, the response was along the lines of we don't publish as we are the only real experts. In fairness, we would say that some of the more senior FBI managers realised that the FBI might have been out of touch with how others viewed hair examination. The proof of this was that they invited outsiders to this landmark meeting that established a Committee on Hair Examination. The preliminary report from that Committee was published in 1985 (Anon, 1985a). Discussions at that 1985 meeting covered all aspects of the hair examination process, including how to report hair findings and appropriate wording. It would be fair to say that opinions differed greatly, and discussion was robust. Our (JR) recollection was that the more senior FBI representatives accepted that the days of claiming hair examinations could uniquely identify an individual were over. However, it was still their firm belief that hair examinations could offer strong support, with discussion focused around how this might be expressed. The RCMP, through Barry Gaudette, were strong contributors to this discussion. RCMP guidelines around this time recommended the use of the term "consistent with".

Evidence given by the RCMP examiner in Driskell (Anon, 2006) was typical of the RCMP as late as the early 1990s.

> when I say that a hair is consistent, as I have in this case, that means that the hairs have all of the features that the known samples have, within normal biological variation, and there's nothing, nothing you would—that you can't account for. So that if there was some feature, for example abnormal colour or something like that, that would cause that hair to be eliminated. So, it falls exactly within the range of the

variation of the known sample with no unaccounted-for differences whatsoever.

And the point about this type of analysis is that it's not a positive identification, all right, because the only way you could do that is to look at all the hairs from all the person's head that exist, and that's an impossibility. But I can tell you, based on my experience, that the chances of just accidentally picking up a hair and having it match to a known sample are very small. So, if the hair is consistent, that means it either came from the same person as that known sample or from somebody else who has hair exactly like that.

And in order to give you a sort of guideline or a rule of thumb to determine how much weight to put on that, you can look around the room and just see how many people even have similar hair styles. If you look at one hair and you examine those 20 features, it's got even more information than you can see by looking at different lengths of hair and different colours and different hair styles. That's not to say that you can't accidentally meet somebody or two people on the street that have exactly the same kind of hair, just like sometimes you accidentally mistake one person for another, but the chances are not very high.

The RCMP guidelines suggested that hair examiners respond in the following way:

> When careful examination by a qualified examiner indicates that a questioned hair is consistent with a known source, there are two possibilities. Either the hair actually originated from that source, or there was a coincidental match. Since it is possible for two different people to have hairs which are indistinguishable by present methods, it is known that coincidental matches can occur in forensic hair comparison. However, based on my knowledge and experience as a hair examiner, I am of the opinion that such coincidental matches are a relatively rare event. The explanation that the questioned hair actually originated from the known source is generally the more likely of the two.

The RCMP approach was based on the view of Barry Gaudette that coincidental hair "matches" were rare. Hence, our point is this. The FBI was not alone in still placing considerable weight on the value of a hair "match". Whilst both FBI and RCMP management accepted that hairs were not absolute identification, they were far from throwing in the "proverbial towel". The result was that for the following decade or more, evidence presented by FBI and RCMP hair examiners had an underlying, or implied, view that hair evidence could exclude all other potential sources because a coincidental "match" was a rare event. Testimony would often include some use of numbers or statements based on "experience" that in x number of years the examiner had only ever seen a very small number of hairs they could not distinguish. Reference was often made to the number of features that matched based on there being about 20 plus such features. Without going into the statistical errors of this approach, this reliance on numbers may be telling about the way in which examiners were trained to think about the process of hair examination and comparisons. Typically, side-by-side images of hairs would also be shown to juries to demonstrate the "match".

Putting to one side any detailed discussion about acceptable terminology, beyond stating that "similar", "consistent with" and "match" should in our firm view be avoided, the approach used by FBI examiners, and others, may be best understood by looking at human expertise. Dror (2011) has written extensively on this subject. He states that "expertise is correctly, but one-sidedly, associated with special abilities and enhanced performance. The other side of expertise is surreptitiously hidden. Along with expertise, performance may be degraded, culminating in a lack of flexibility and error".

Dror (2011) focuses on fingerprint expertise. However, we would argue that the common element that links feature-comparison methods, is an over-reliance on the idea of a "match" based on a magical number of points of similarity. The training of pattern recognition experts is aimed at producing an expert that can "apply automatic sequences quickly and efficiently", making sense of "signals and patterns" and dealing "with low quality or quantity of data" (Dror, 2011). According to Dror, experts report that they "see things differently", the implication being that as a "non-expert" one cannot be expected to understand how the expert reaches a decision. The problem with this approach is that if the "expert" cannot explain how they reached a decision to the fact finder (jury), then the invitation to the fact finder is "trust me, I'm an expert!" Dror (2016) describes how experts consolidate and integrate complex sequences of steps into a routine as follows:

> By chunking steps together into a single entity or action, the experts not only achieve quick performance in terms of execution time, but they are able to do more because these processes are more computationally efficient. Such mental representations and information processing many times give rise to automatization. Experts rely on such processes especially in domains that require complex decisions and actions under time pressure and risk. Automaticity is so efficient that many times it does not require conscious initiation or control, and it may even occur without awareness.

As Dror (2016) goes on to point out the downside of *automaticity*, "the expert cannot fully account and explain, or even recall their actions". Expert knowledge becomes inaccessible, does not get published, is not subjected to peer review processes and training is compromised. This was very much the situation in the FBI of the 1980s with respect to hair examinations!

Dror points out the potential psychological negatives that some experts exhibit, including over-confidence, refusal to listen to others or take advice or pay sufficient attention to detail.

It is also important to understand the cognitive profile that best meets the abilities required to perform the task (Kelty *et al*, 2009). As Herrington and Colvin (2015) point out, command and control will remain an important element of Policing, but as organisations grapple with increasingly complex problems, this will require innovation and experimentation. Whilst these authors were referring to Police leadership, they could just as well have been talking about complex forensic tasks such as fingerprint examination or hair examination. Given the highly structured command and control environment in which the FBI hair examiners worked in the 1980s and 1990s is it a surprise that their approach to hair examination followed a traditional policing management style?

To summarise, to the best of our knowledge hair examiners in the FBI are no longer agents, and all now have appropriate scientific qualifications. Hence, the cultural and cognitively derived philosophical issues that were the underlying reasons explaining the approach to features-based examinations in the FBI, and others at that time and before, should now be a thing of the past, at least for hair examiners. Although not a universal panacea to solve all problems, quality systems and accreditation have placed most forensic laboratories on a sounder footing.

1.5 WHERE TO NOW—FINAL CONCLUSIONS AND THE WAY FORWARD

The purpose of this introductory chapter has been to consider the history of hair examination by looking at how it has evolved over the last hundred years and to understand where hair examination lies today and what still needs to be done if the examination of hairs is to remain a core forensic expertise.

In 1999, Robertson wrote that forensic hair examination was at a crossroad and that its future would be determined by the dual impact of Daubert and DNA (Robertson, 1999b). To some extent, Daubert is old news but issues of scientific validity continue to be raised through major reports such as the National Academy of Science (NAS) report and the PCAST report. The final outcomes of the enquiry into former FBI practices in reporting hairs have yet to be made public, if indeed they ever are made public, and of course, the issues that have already been identified through this enquiry are not necessarily limited to the FBI only.

However, we would argue that although there may still be issues to consider from this enquiry, this is also old news.

In the context of addressing the PCAST questions of scientific validity and validity as applied, we believe that when hairs are viewed as a biological material, yielding DNA suitable for analysis, then the scientific validity that is relevant is that of DNA analysis, interpretation and reporting. Except for the interpretation of ultra-trace or touch DNA there is general acceptance that DNA is the one area where scientific validity is not in question. As with any aspect of scientific analysis the "validity as applied" of DNA is dependent on it being applied to appropriate standards. We have intentionally chosen to not include a specific chapter in this book on DNA analysis of hairs and have limited our consideration of DNA to issues specific to hairs as a source of DNA. In Chapters 5 and 6 of this book we will look in detail at the interpretation and reporting of findings from hair examinations, and in Chapter 7 we will present what we believe is required regarding training to ensure compliance with best practice in these areas.

The focus on hairs as a biological source of DNA has been a paradigm shift for the examination of hairs within contemporary forensic practice, but it has **not** removed the need for the appropriate and proportional use of microscopic examination of hairs. In Chapter 3 we will discuss recognition, recording and recovery considerations for hairs within a contemporary case management context based on a triage model aimed at the early identification of hairs suitable for nu-DNA analysis and elimination of hairs that clearly have no potential evidential value and minimising type 1 and type 2 errors. We will also take a broader criminalistics view of the information value of hairs in answering case- or activity-level questions as well as being a valuable source of biometric or identification information. In this regard we will attempt to redress the balance in the current minimalist approach all too often seen in many forensic laboratories with respect to the examination of hairs.

CHAPTER 2

Fundamentals of the Biology and Chemistry of Hairs

2.1 INTRODUCTION

Living things grow, OK? Hair grows, OK? So, hair must be alive, OK? No, came the answer, hairs are not alive, it's dead! Well, I said, how does hair grow, then, if it's dead?

(Hackett, 1984)

As Harding and Rogers (1999) state the answer to this paradox is that the hair root embedded in the skin is a living tissue and grows to produce the hair we see, which is dead tissue, or as Montagna (1963) puts it "as dead as a rope". The living part of a hair develops as an invagination of the epidermis and, hence, is an appendage of the skin.

The hair follicle is of interest to cell and molecular biologists due to the very high rates of cell division that make it an excellent model to study cellular processes in animals. Indeed, most hair research uses non-human animal hairs and much of what we know about the biology and chemistry of hairs comes from models derived from animal studies. Notwithstanding, the focus of this book is on the forensic examination of human hairs and we will only draw on animal hair studies as they are needed to make sense of the forensic examination of human hairs.

A key aspect of the biology of hairs is their growth cycle as this explains the scientific basis for hair loss that is the basis of why we see hairs in the forensic context. Hence, a major focus in this chapter will be understanding hair growth and development. Here it is important to point out that in most animal species, follicles and growth cycles are synchronised and hair loss is seasonal. However, for humans each follicle and its growth cycle are independent. In normal growth humans do

DOI: 10.4324/9781315210650-2

not lose hairs as groups although, as we will see, that does not mean that several hairs may not be removed at one time due to grooming activities or the use of force. Hence, for humans as each hair grows independently of all others, they should be considered unique and no two hairs will be exactly alike.

Relative to most animal species humans are essentially hairless, or glabrous. Whilst, we have about the same number of hair follicles as our closest relatives, the apes or primates, as we will discuss later in this chapter, in humans not all follicles produce terminal hairs, hence the appearance of being less hairy! Without going deeply into the theories as to why humans have become the so-called "naked ape" it is fair to say that hairs no longer have a primary role for humans in regulating temperature, protecting us against the elements, as camouflage or ensuring we can float in water—all useful roles for hairs in other animals. Hair in humans has become largely cosmetic and ornamental. For those interested in the broader socio-cultural aspects of hair the reader may be interested in a little book simply called "Hair" (Boccalatte and Jones, 2009). Jablonski (2010), in discussing the possible origins of human hairlessness, explains the differences in hair profile in humans based on tightly curled hair protecting the top of the head from excess heat. This is due to the role of hair in creating a barrier layer of air between the scalp and the hot surface of the hair. Hence, on a hot day the hair absorbs heat whilst the barrier layer remains cooler. Tightly curled hair provides the most effective barrier as it increases the thickness of the space between the surface of the hair and the scalp, allowing air to blow through. Humans may have developed straighter hair as they migrated to cooler climates. Interestingly, there is also evidence that Africans have less dense hair coverage than Caucasian individuals (Loussouarn, 2001).

As Harding and Rogers (1999) state "hair seems to be a dilemma … if we've got it, it's in the wrong place, or it's the wrong colour or the wrong style. If we haven't got it, we wish we did, or if we have lost it, we try to make it look as though we still have it. We worry about the ones that fall out, and the ones that turn grey, or worse yet, white!"

Humans attempt to compensate for all the above through grooming and treatments aimed at stopping hairs from falling out, removing unwanted hairs, replacing hairs/follicles from one location to another or changing the colour or other physical features of our natural hair. All this adds a rich source of acquired and non-inherited features to the potential armoury of the forensic hair examiner.

We also know that the biology of hairs is under genetic control and, hence, there is a genetic inherited basis underpinning hair features.

And of course, the underpinning premise for hairs as a physical material that we encounter in forensic examinations, is that they are shed and are available to transfer and persist in all sorts of situations, providing a potentially rich source of information that can assist investigations with biometric and criminalistics questions of interest.

In this chapter, we will lay down the foundational knowledge to understand the biological and chemical basis for understanding how hairs grow, die and are removed (or lost) and the biological basis for variation that exists between hairs from one person and variation in hairs from different people.

2.2 HOW IS THE FIRST HAIR FOLLICLE FORMED?

The hair follicle is a downgrowth or invagination of the epidermis into the dermis. Ebling (1980) describes the development of the hair follicle as comprising five stages, *pre-germ, hair germ, hair peg, bulbous hair peg and hair follicle*. These are illustrated in Figure 2.1 and are relatively self-explanatory. For a detailed description of these stages see Harding and Rogers (1999).

An important stage in the development of the hair follicle is the bulbous peg as this is when a solid swelling of cells, called the **bulge**, develops from an outgrowth of the peg. The bulge is also where **the arrector pili muscle** becomes attached.

Above the bulge is the point in the follicle where all follicular components persist throughout subsequent hair growth cycles. In the mature hair follicle, it is below this point the parts are resorbed as the hair stops growing (**catagen** growth phase) and moves to its resting or **telogen** growth phase. A second swelling above the bulge forms the lipid rich **sebaceous gland**. The sebaceous glad is formed around the six-month stage of foetal growth.

Around the same time in foetal development the **germinative cells** in the hair bulb adjacent to the dermal papilla begin to actively divide. The first cells created are the inner root sheath (IRS) cells that align themselves along the follicle and form the **hair cone** above the dermal papilla. Pushed up by the dividing cells from below the hair cone forces its way upwards between the cells of the central core of the hair canal (Harding and Rogers, 1999). When about halfway up the follicle the hair starts to harden and the tip of the hair starts to keratinise. Cell differentiation starts to produce the various elements of the hair and the hair breaks through the epidermis.

The first hairs formed are very fine, have no medulla and are called **lanugo hairs.** Lanugo hairs are shed *in utero* around the 7th and 8th month

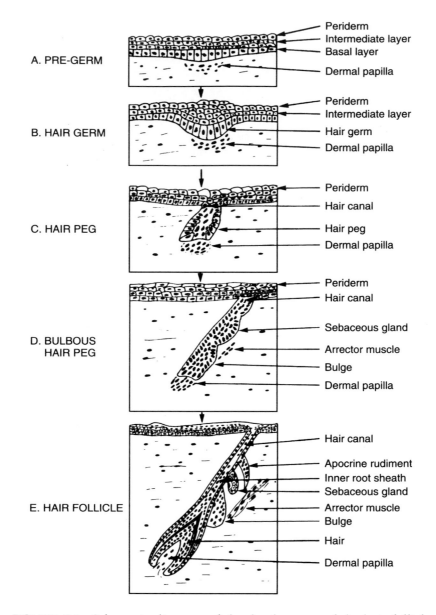

FIGURE 2.1   Schematic diagram of the development of the hair follicle in the foetus. (Source: Ebling (1980); *used with permission.*)

of gestation and replaced by either new lanugo hairs or **vellus hairs**. However, this change can continue into the first few months after birth and, hence, at birth the follicles can be at all stages of the hair cycle (Montagna and Parakkal, 1974).

Active growing hairs are called **anagen** hairs.

2.3 THE ROLE OF THE HAIR FOLLICLE IN THE GROWTH OF HAIR FIBRES

Following the development of the original hair follicle, follicles will produce hairs throughout the life of the human following a pattern of growth, loss and regeneration called the **hair cycle** (Chase, 1965). This cycle is a dynamic and continuous process that involves a growth phase, **anagen**, a transition phase, **catagen** and a resting phase, **telogen**.

Figure 2.2 provides a schematic representation of a cross section of a mature actively growing hair follicle showing the main features and regions where the main events of the hair cycle are enacted.

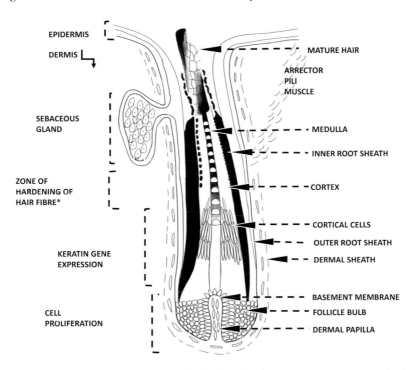

FIGURE 2.2   Schematic of hair follicle. The hair unit is composed of a hair follicle, arrector pili muscle and sebaceous gland. Hair follicles vary in their size and the type of hair present, but all have a similar structure showing various main features and the regions where the main events of cell proliferation and keratinization take place. The cells in the bulb region move into different streams that form into the cortex, the cuticle, the medulla and the inner root sheath. (*Original diagram by Dino Todorovic, scientific illustrator, modified from* Powell, BC & Rogers, GE 1997, 'The role of keratin proteins and their genes in the growth, structure and properties of hair', in Jolles, P, Zahn, H & Hocker, H (eds), *Formation and Structure of Human Hair*, Switzerland: Birkhauser Verlag.)

It is important to remember that the hair follicle is three dimensional and comprises a series of roughly cylindrical and concentric layers. The most external layer is the **outer root sheath** (ORS) that is next to the epidermis. Next is the IRS that encloses the hair and physically holds the growing hair in the follicle. The hair itself has at least two cell types and sometimes a third cell type. The first cell type is the outer surface layer or **cuticle,** consisting of a series of flattened, overlapping (or imbricate) layers of cells. The cuticle encloses the second cell type of elongated spindle shaped cells found in the central core or **cortex.** The hair may also include a third cell type forming the **medulla.** When present, the medulla is in the centre of the hair shaft and, in human hair, consists of amorphous or unstructured cells.

With the exception of the ORS, all the above cells types arise from germinative cells in the follicle bulb and proliferate during the anagen growth phase of the hair cycle. These germinative cells are separated from the **dermal papilla** by a basement membrane. The dermal papilla is a projection of tissue into the base of the hair follicle and controls the physical characteristics of the hair as well as hair growth. The dermal papilla is the only permanent part of the hair follicle and remains through the hair cycle as a condensed ball of cells at the base of the shortened follicle throughout the telogen growth phase (Oliver and Jahoda, 1981). A mature hair follicle is about 4–5 mm in length (Kligman, 1962).

As discussed in Section 2.2, attached to the mature follicle above the bulge is the arrector pili muscle and one or more sebaceous glands. The combined follicle and glands are referred to as the **pilosebaceous unit** (Pinkus, 1958). The area above the bulge is referred to as the **upper follicle** or **permanent zone.** The region of the follicle below the bulge is called the **lower follicle** or **transient zone** as this is the part of the follicle that regresses during the catagen phase of the hair cycle.

As previously stated, the germinative cells are amongst the most actively dividing cells in the human body, requiring very active metabolic activity in the region of the hair follicle. For example, there is a four-fold increase in biochemical activity (pentose cycle) doubling the rate of adenosine triphosphate (ATP) synthesis and producing the substances responsible for fatty acid and nucleic acid metabolism (Adachi and Uno, 1969). Furthermore, because of the highly active cell division of the germinative cells anagen hairs have high levels of nu-DNA with a single root reported to yield as much as 50 ng (von Beroldingen *et al,* 1989). In Chapters 1 and 4 we discuss hairs in more detail as a potential source of DNA for forensic analysis. The germinative cells also include cells responsible for pigment formation in hairs. These cells, called **melanocytes,** are present around the apex of the dermal papilla. Each melanocyte has several dendrites that extend upwards between the presumptive cortical cells (Swift, 1977). We will return to consider the role of melanocytes and hair colour in Section 2.7.

Viewed from a functional perspective the follicle can be considered to comprise four zones as illustrated in Figure 2.2. These zones are **cell proliferation and differentiation, keratin gene expression, keratogenous or hardening,** and the zone of **IRS degradation.**

In zone 1, cell proliferation and differentiation results in up to six continuously dividing distinct cell streams that move upwards and outwards to form the tissues of the follicle and the hair fibre. The most central cells give rise to the medulla (when present) then progressing outwards, the next layers give rise to the cells of the cortex, the cuticle and then the layers of the IRS.

Harding and Rogers (1999) describe the cellular development stages that occur as cells differentiate into those comprising the medulla, cortex and cuticle.

Medulla cells arise from amorphous granules that coalesce and create a hardened, highly insoluble, citrulline rich protein. In human hairs the medulla has an amorphous or unstructured appearance when viewed with transmitted light microscopy (TLM). For other animal hairs the medulla often displays a well organised and structured appearance at the TLM level.

Differentiation and development of cortical cells in human hair has been described by Birbeck and Mercer (1957). An important feature in the early stages of the development of cortical cells is that there are intercellular gaps of 2–3 um present as the cells move past the melanocytes in the dermal papilla and it is these gaps that allow the dendrites of the melanocytes to penetrate between the developing cortical cells. The melanocytes enter the cortical cells by a phagocytic process and pigment granules are then released by partial digestion inside the cortical cells (Birbeck and Mercer, 1957; Swift, 1977). As the cortical cells develop, they change from being round to elongate, spindle shaped cells arranged parallel to the axis of the follicle. The protein in the cells becomes filamentous and eventually form microfilaments and macrofilaments (now called **keratin intermediate filaments**). The pigment granules are arranged in rows between the macrofilaments.

As the hair fibre moves up the follicle into the zone of keratin gene expression the whole cortical cell is filled with microfilaments and macrofilaments. The now distorted and spindle shaped nucleus of each cortical cell is in the centre of the cell surrounded by keratin-based filaments. As the cortical cells further develop and move from zone 2 to zone 3 (keratogenous or hardening) there are also changes in the chemistry of the keratin. This is the result of disulphide bond cross linking of the sulphur rich proteins as the cortical cells dehydrate and harden. It has been estimated that the diameter of the hair fibre is reduced by about

25% at this terminal stage in the development of cortical cells (Rook and Dawber, 1982). The cortex is separated at this stage from the cuticle by the development of the **cell membrane complex** (CMC). The hair fibre is biologically inactive at this stage and is functionally "dead".

Cuticle cells arise from a single layer of germinative tissues outside those that form the cortex. As the cells develop and mature, they change shape becoming elongated and flattened. They then tilt to the side over each other to produce a layer of overlapping cells (Swift, 1977).

Finally, the IRS comprises three layers of cells, respectively, the cuticle layer, Huxley layer and the Henle layer. These three layers arise as separate but adjacent concentric cylindrical layers of germinative cells in the hair bulb. Harding and Rogers (1999) describe the development of the IRS in detail. The important aspect of the IRS is that as it moves up the follicle its cells mature and harden low down in the follicle forming a sheath that moulds, shapes and supports the forming hair fibre. This support continues up to the level of the sebaceous gland at which point the IRS cells dissociate and fragment (Fraser, 1969; Straile, 1965). As previously mentioned, the IRS also holds the hair fibre in the follicle through the interlocking action of the IRS cuticle cells which have downward pointing scales that interlock with upwards pointing scales of the hair cuticle. When a growing hair is "plucked" the IRS will sometimes be removed and is seen at TLM level as a translucent layer attached to the lower part of the hair fibre.

For a detailed treatment of the cell biology and development of the ORS, dermal papilla, sebaceous glands and the role of blood capillaries, nerves and muscles in the dermis the reader is referred to Harding and Rogers (1999).

2.4 THE HAIR CYCLE

For the purposes of description, the hair growth cycle is broken down into three phases, **anagen, catagen** and **telogen**. Anagen is the growing phase, catagen a transition phase between anagen and telogen, and telogen the resting phase. The telogen phase is also sometimes referred to as the mature or quiescent phase. Figure 2.3 shows a diagrammatic representation of the hair follicle at the three different stages of the growth cycle; Figure 2.4 shows the appearance of these phases in longitudinal section (Harding and Rogers, 1999); and Figure 2.5 shows the appearance of these phases as seen by both low-power microscopic examination (LPM) and TLM. However, it is important to reiterate that the hair growth cycle is a dynamic and continuous process.

In Section 2.3 we have already described the formation of the "first" hairs generated from the hair follicle. The formation of an anagen hair takes about three months for the hair to reach the surface of the epidermis

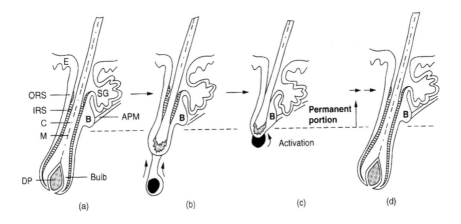

FIGURE 2.3   Diagrammatic representation of hair follicles at the different stages of the growth cycle. Anagen (a) is the active growth phase during which follicle development takes place and the hair fibre is produced. Catagen (b) is the regression phase in which tissue changes occur as the follicle approaches telogen (c), the resting phase. The next anagen is thought to be initiated by the interaction between the dermal papilla cells and the pluripotent stem cells located in the bulb region. E, epidermis; ORS, outer root sheath; IRS, inner root sheath B, bulge; APM, arrector pili muscle; SG, sebaceous gland; C, hair cortex and cuticle; M, medulla; DP, dermal papilla. (From Harding and Rogers (1999), modified from Cotsarelis et al (1990), copyright cell press, *reproduced with permission.*)

after which the hair will continue to grow for a period of at least two years and as long as six years or more (Robbins, 1988). Typically, a human scalp hair will grow at a rate of about one centimetre per month. Hence, over a six-year growth period in theory the maximum length of a scalp hair should not exceed 72 cms. As there are numerous examples of humans having scalp hairs longer than 72 cm, it is clear that hair growth rates and the length of the anagen growth phase are only estimates and that in reality hair can reach much longer lengths than the theory would predict.

As we have commented on previously anagen hairs are firmly held in the follicle by the IRS. When an anagen hair is removed quickly the root end may have attached elements of the lower papilla and/or sheath tissue, although this can vary between individuals (King et al, 1982). As anagen hairs are actively growing, hairs that are coloured will have pigment right the way down to the hair root.

The catagen growth phase is defined by a gradual, orderly period during which growth shuts down over a one to four-week period. During this time the hair follicle regresses back to the area of the bulge or the permanent

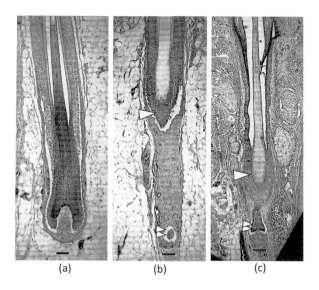

(a)   (b)   (c)

FIGURE 2.4 Hair growth cycle. Transmission electron micrographs of longitudinal sections of human scalp hair follicles showing different stages of the growth cycle. Note that some separation of the tissue layers has occurred during section preparation. (a) Follicle at anagen. (b) Follicle at catagen. The club shaped root has formed (arrowhead) and the follicle has regressed, leaving the dermal papilla at its original depth (double arrows). (c) Follicle at telogen. The club shaped root (arrowhead) has minimal connection with the follicle. Bar = 100 μm. (From Harding and Rogers (1999), copyright, Taylor & Francis Group, *reproduced with permission*.)

portion of the hair follicle. Cell division decreases and eventually stops, and the IRS disintegrate and melanin production ceases. The appearance of the root end becomes brush like, and these hairs are quite loosely held in the remaining hair follicle. If catagen hairs are pulled or removed through grooming (brushing or combing) they may become detached with remnant cellular material of the follicle, referred to as a *follicular tag*.

In the telogen growth phase hairs are no longer growing as the cells in the lower region of the follicle are no longer active. The telogen follicle is about one third the length of the anagen follicle and held in the follicle only by the club shaped root. Nonetheless, telogen hairs do not normally just fall out! Telogen hairs are removed due to grooming activities such as combing, brushing, washing or drying with a towel. If not physically detached eventually the resting hair will be pushed out when the follicle is reactivated, and a new hair is produced. Telogen hairs are easily recognised due to the presence of a club shaped root, usually with no pigmentation or medulla near the root end and often the presence of

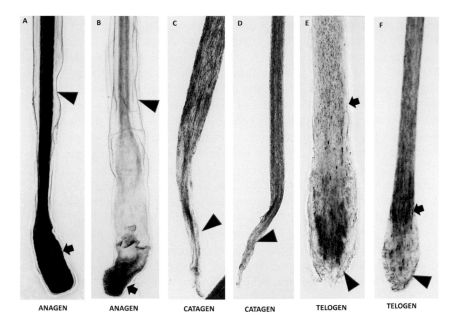

| ANAGEN | ANAGEN | CATAGEN | CATAGEN | TELOGEN | TELOGEN |

FIGURE 2.5 Light micrographs of the three root growth phases of human hair. (a) & (b) Both images are of anagen roots (arrow); (a) from a black-haired person and (b) from a light brown-haired person. In both hairs the sheath is obvious (arrow heads) and the dark portion of the root in (b) are the newly formed cortical cells with pigmentation (arrow). (c) & (d) These two hairs show a classical catagen root type (arrowhead); sheath material is absent and pigmentation and cell division has ceased. (e) & (f) Classical "cotton bud" or "club" shaped roots are typically telogen stage growth (arrowheads). There is remnant pigment granulation in the club end of (e), and a zone of no pigmentation above the root (arrow). Cortical fusi are often observed in this zone. Hair (f) is in an earlier telogen stage than hair (e) and does not yet have a pigment free zone (arrow). Neither hair ([e] & [f]) has any sheath material.

cortical fusi. The telogen phase typically lasts for 3–4 months, during this time a new anagen phase commences. This involves the regeneration of an almost entirely new follicle in a process like the initial development of the follicle as described in Section 2.2.

As previously discussed, the growth cycles of human hairs are not synchronised, and the growth cycle of each hair is independent of all other hairs. This results in a growth pattern called *scattered mosaic*. Hairs are lost or shed on a regular and even basis at a rate for scalp hairs of 100–150 per day. This is based on there being between 100,000 and 150,000 follicles in the scalp, 10% of these being in the telogen growth phase with a telogen period of around 100 days and "available" to be shed (Orentreich, 1969). Estimates of the percentage of scalp hairs in anagen, catagen and telogen are quite varied but generally around 90% will be in anagen growth phase, only 1–4% in catagen and 10% or slightly less in telogen. Montagna (1976) has commented that there are too many differences at the level of individuals to place any great weight on estimates such as those quoted above. For example, males tend to have a higher number of telogen hairs than females, and this increases with age (Barman *et al*, 1969). It has also been shown that the percentage of hairs in the anagen growth phase increase during pregnancy and shedding rates drop to 10–15 hairs per day (Zviak and Dawber, 1986). There is a genuine scientific basis for pregnant women believing that their hair is thicker during pregnancy! Of course, the downside of the latter is that following pregnancy there will be more hairs available to be shed.

The percentage of follicles in the three growth phases and the length of time for each phase also varies across hairs from the different areas of the body (Montagna and Parakkal, 1974).

The forensic significance of the dynamics of the hair cycle is that unless hair has been removed with some force, most hairs recovered during forensic examinations will be telogen hairs. It has been estimated that about 95% of hairs recovered in case work are in the telogen growth phase.

2.5 HAIR DISTRIBUTION, TYPES AND GROWTH RATES

2.5.1 Distribution

It is estimated that at birth humans have between 2 and 3 million hair follicles, out of which about 100,000–150,000 are in the scalp. Although there is some evidence of limited development of additional follicles after birth (Pinkus, 1958), generally humans are born with all the hair follicles they will ever have in their lifetime.

The remaining follicles are in the face (10,000–20,000), the trunk (425,000), the arms (220,000) and the legs (370,000). Clearly when combined these estimates of follicle numbers do not add up to 2–3 million. As it is simply not possible to count the total number of follicles in an individual, estimates are made by counting small areas with relatively low numbers counted and then extrapolations made from these numbers. Hence, total follicle numbers should be treated only as broad approximations. There is also wide variation in the density of hair follicles across different body areas with typical densities in the scalp of 200–300 follicles per cm^2 to only 6–31 hairs per cm^2 for pubic hairs (Astore *et al*, 1979). There are no significant differences in follicle density between females and males. There is evidence that there can be variation in follicle density based on ethnicity with hair density in Africans being lower than in Caucasians (Loussouarn, 2001).

2.5.2 Types

As previously stated, all the follicles in the human foetus produce **lanugo** hairs. These hairs begin to shed *in utero* around the seventh and eight months of pregnancy when they are replaced by new hairs from the same follicles. Most of these new hairs are very short (around 1 mm), fine (4 µm or less) and essentially non-pigmented **vellus** hairs (Robbins, 1988). The third type of hairs are called **terminal** hairs and initially are found only in the eyebrows and eyelashes of humans. Vellus hairs in the scalp are replaced by terminal hairs in first few months after birth although this timing is quite variable. Terminal hairs replace other vellus hairs on the body at different times largely dependent on the sexual maturation of humans. Thus, terminal hairs are sometimes called *primary* or *secondary* hairs or *asexual* and *sexual* hairs. Secondary or sexual terminal hairs develop at puberty and replace vellus hairs in the axilla regions, pubic region and abdomen of males and females and in the facial hairs of males. Although there is some replacement of vellus hairs with terminal hairs pre-puberty in other areas of the body, such as the forearm and legs, this increases after puberty. The extent of replacement of "body hairs" differs between males and females. Typically, about 90% of vellus hairs in males will become terminal hairs with only around 35% in females (Montagna, 1976). It is also important to understand that intermediate types of hair exist, for example, there is no definitive line of distinction between pubic and abdomen hairs.

As humans age some follicles that are producing terminal hairs can revert to producing vellus hairs. This is most obvious in the development of typical male pattern baldness. Some follicles that have produced vellus hairs for a long period can start to produce short terminal hairs later in life such as follicles in the ear and nose of adults as they age, hence, the earlier comment by Harding and Rogers (1999) about having hairs in the wrong place!

As the development of secondary or sexual terminal hairs is a response to changing levels of androgen hormones, and natural levels of these vary between individuals of the same sex, there is also considerable variation in how many follicles will move from producing vellus hairs to terminal hairs, and, hence, how "hairy" an individual may visually appear!

2.5.3 Growth Rates

There have been remarkably few detailed studies of the growth rates of human hair and none that were conducted specifically with a forensic application as their primary objective. Methods used in these studies to estimate growth are also very variable with few measuring the growth of single hairs. From these few studies it is also evident that there is considerable variation in hair growth between individuals. It is also clear that growth rates are affected by a variety of factors including age, sex, ethnicity, nutrition, hormone levels as well as body location. For example, Sims (1967) report hair growth rates for severely malnourished children of almost half that for healthy children.

Hence, from a forensic perspective it is advisable to take a conservative approach to quoting rates of hair growth. What seems to be on solid ground is that human scalp hair for an adult grows at about a minimum of 1 cm per month based on a daily rate of 0.35 mm (Myers and Hamilton, 1951). Myers and Hamilton (1951) report higher rates of growth of scalp hairs pre-puberty with a daily rate of 0.41 mm. Saitoh *et al* (1969) give a slightly higher daily rate of 0.44 mm and Chase and Silver (1969) the highest at 0.50 mm per day.

Growth rates for scalp hair are slightly higher for females than in males (Myers and Hamilton, 1951). Other factors such as shaving, diurnal variation and menstrual cycle in females do not appear to have a significant effect on hair growth rates (Saitoh *et al*, 1969).

Growth rates vary during adult life with a maximum growth rate in the age group 50–69 (Pelfini *et al*, 1969).

Pelfini *et al* (1969) quote daily growth rates of 0.29 mm for thigh hair, 0.35 mm for scalp, 0.36 mm for axilla and 0.40 mm for pubic hair. Myers and Hamilton (1951) give daily growth rates of 0.16 for eyebrow, 0.20 for thigh, 0.30 for axilla, 0.35 for scalp and 0.38 for beard hair. Hence, overall, it would appear that shorter hairs such as eyebrow and thigh (body) hairs have lower daily growth rates although beard hair is similar if not higher than some estimates of scalp hair growth.

In summary, forensic scientists should be conservative in estimating growth rates based on the limited data available.

2.6 MORPHOLOGY AND ANATOMY

2.6.1 Morphology

The shape of human hairs, or morphology, and the potential length of uncut hairs varies because of body location and the ethnicity of the individual. Hicks (1977) produced summary tables of the features of hairs based on body location and ethnicity. These are discussed in some detail in Chapter 4.

These descriptions should only be considered indicative for two reasons. First, the demarcation between body locations is not always distinct and there is some transition between areas. Second, with respect to ethnicity, the concept of distinct ethnic or racial groups is now somewhat

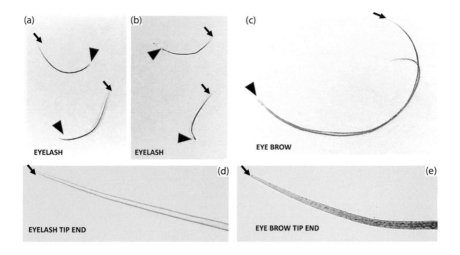

FIGURE 2.6   Somatic hair shaft profiles—Image Set A. The LPM images (a, b, c) show the distinctive sabre shape associated with most eyelashes and generally most eyebrows. The arrowheads indicate the root ends and the arrows the tip ends. In the TLM micrographs, (d, e) the almost clear tip end of the eyelash is typical (mascara is used to darken these), while the eyebrow tip end depicted here has more colour. (*Continued*)

outdated. We live in an increasingly multicultural world where phenotype mixing is more common than in the past.

With respect to the general descriptions relating to body area determination these are based on Caucasian hairs. For example, scalp hairs are described as "long with moderate shaft diameter". This general description must be somewhat modified when ethnicity is then considered. Hair shaft shape for African scalp hairs is described as showing "prominent twist and curl". Cross sectional shape of scalp hairs is also dependent on ethnicity with Caucasian scalp hairs being oval, African scalp hairs flattened and Asian scalp hairs round.

Figure 2.6 shows several examples of hairs of different somatic origin.

2.6.2 Anatomy

The human hair has three distinct cellular components these being the **cuticle**, the **cortex** and the **medulla**. The first two components are always present, and the medulla may or may not be present and/or readily visible. Figure 2.7 shows a diagrammatic representation of these three components each of which will now be considered in more detail.

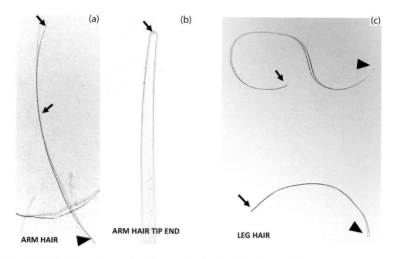

FIGURE 2.6 (*Continued*) Somatic hair shaft profiles—Image Set B. The full-length arm (a, b) and leg hairs (c) shown here are from the same male person—the root ends indicated by arrowheads and the tip ends by arrows. Female arm and leg hairs are often finer and lighter in colour though not always. However, the almost clear/white tip end from the arm hair is typical, noting the abraded tip end that is caused by rubbing against clothing. (*Continued*)

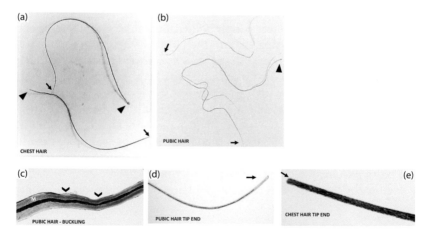

FIGURE 2.6 (*Continued*) Somatic hair shaft profiles—Image Set C. Full-length chest hair (a) and pubic hair (b) from male persons showing different shaft profiles. In many cases for males, the two hair types may appear the same—both buckled (c) and curly or curved and not curly/buckled. Arrowheads indicate the root ends while the arrows point to the tip ends of the hairs. Similarly, to the arm and leg hairs, the TLM micrographs show rounded/abraded (d, e) tip ends of both hairs caused by rubbing against clothing. Buckling, (chevrons) seen in the shaft of the public hair (c) allows the hair to curl and twist. This pubic hair has a very prominent medulla (M).

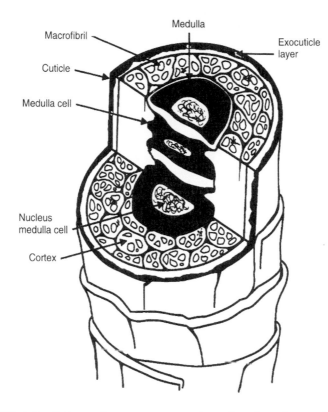

FIGURE 2.7 Schematic cut-away section showing the major cellular structures, cortex, medulla cells separated by air spaces, nuclear remnants in the medulla cells, macrofibrils – the aggregates containing the filaments and matrix. (From Harding and Rogers (1999), copyright Taylor & Francis Group, *reproduced with permission.*)

2.6.2.1 Cuticle

As we have previously briefly described the outer layer of the hair shaft, the **cuticle**, is comprised of overlapping (imbricate) layers of scales. The scales overlap laterally and longitudinally to encircle and enclose the cortex. The scale edges slope outwards pointing to the hair tip. Individual scales are approximately rectangular, about 50–60 μm long and about 0.5 μm thick (Swift, 1981). Due to the scales overlapping the part of the scale that is visible is about one sixth of the length of an individual scale. Overall, the cuticle is effectively a multilayer of six scales with an overall thickness of 3–5 μm. The scales or cuticle cells have been shown to have three layers, the inner *endocuticle*, the outer *exocuticle* and a very thin layer called the *A layer* on the outer edge of the exocuticle. From a purely forensic context these layers are not visible at light microscopic levels of resolution and are only visualised with electron microscopy. Harding and Rogers (1999) describe the ultrastructure of the cuticle in detail. The whole cuticle is surrounded by very thin layer (10 nm), the *epicuticle*, providing a hydrophobic or water repellent outer surface to the hair shaft. This layer helps protect the cuticle from damage and assists hairs remove dirt due to the phenomena called "directional friction" (Swift, 1977). Directional friction is where the friction is less going in the direction of the root to the tip than the opposite direction. The epicuticle is resistant to enzymes and chemicals due to its highly cross-linked protein (Zahn *et al*, 1994). Hydrophobicity is due to a monomolecular layer of a C21 saturated fatty acid, 18-methyl-eicosanoic acid (18-MEA) (Evans *et al*, 1985). 18-MEA also has a role in directional friction.

When viewed at a TLM level, under appropriate conditions, or after taking a scale cast, the overall pattern of scales in human hairs does not vary within or between individuals. When viewed along the length of the hair shaft the scale edges may start to change from a smooth edge to a more irregular pattern typically seen in human scalp hairs. This is due to damage caused by grooming, washing, weathering and other wear and tear. Figure 2.8 shows the typical irregular imbricate pattern of human hairs along with some examples of damaged hairs. Taken to its extreme damage may be so severe that the cuticle is removed exposing the inner cortical cells seen as "split ends" or "brush ends". Examples of tip end damage to hairs are shown in Figure 2.9.

In summary, although a vital component of hair anatomy, in a forensic context there are no inherited cuticle features that are useful to

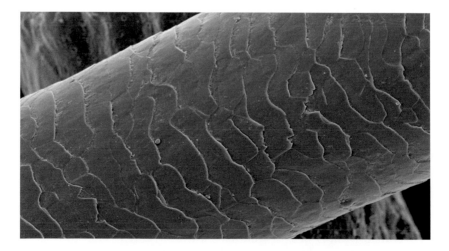

FIGURE 2.8   Scanning electron micrograph showing the imbricate cuticle scale pattern of human hair.

differentiate human hairs. The situation for animal hairs is an entirely different story that will be considered in Chapter 3.

2.6.2.2 Cortex

The **cortex** in human hairs comprises the major component of the hair shaft protected and enclosed by the cuticle. Cortical cells are elongate, spindle shaped (fusiform) and are aligned along the lengthwise axis of the hair shaft. The cells are held together by the CMC. Individual cortical cells have an internal structure of keratin microfibrils or keratin intermediate filaments (keratin IF) embedded in a matrix of sulphur-rich proteins, keratin-associated proteins (KAP) (Fraser *et al*, 1972). Harding and Rogers (1999) discuss the molecular substructure of hair proteins in detail.

The disulphide cross linking of these sulphur rich proteins imparts tensile strength to hairs and makes them largely insoluble. The latter has implications for some cosmetic treatments of hairs such as permanent dying where the disulphide bonds are broken and then reformed. As not all bonds will be re-established this can result in a weakening of the hair shaft which can result in the cortical cells becoming disconnected. This

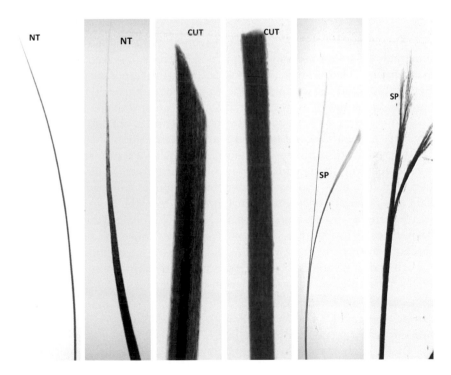

FIGURE 2.9 (*Continued*)

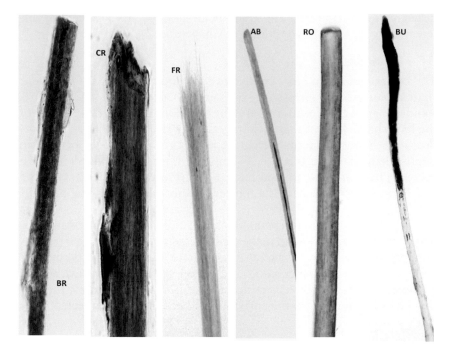

FIGURE 2.9 (*Continued*) Hair tip features. These light micrographs illustrate both common and less common features observed during hair examination. The common ones include the natural taper seen in hair that has not been cut; cut (cut) where the hair has been barbered or cut at the hairdresser; split (SP) as in "split ends" that occurs as the hair is exposed to environmental damage and daily grooming and is not cut; frayed (FR) where the hair ends become brush like in appearance due to the same factors causing splits and sometimes chemical permanent waving; abraded (AB) where the natural tapered hair tip is rubbed or smoothed due to interaction with clothing—especially occurs in body/pubic hairs; rounded (RO) hair tips generally occur between haircuts where the sharper cut ends weather and wear to a more RO shape; burnt (BU) where the person has been exposed to heat/fire significant enough to cause the hair tips and shafts to burn or be heat affected; and finally, less commonly seen are the crushed (CR) and/or broken (BR) tip ends resulting from mechanical forces impacting on the hair/head as in head injuries.

can be seen at the TLM level as *coarse cortical texture* or, if the hair is broken at the tip end, as obvious fibrils.

In the 1980s, there was considerable interest in the potential for hairs to be individualised based on differences in the proteins of hairs. Promising as this research may have been, its potential value became redundant as DNA testing developed.

As was discussed in Chapter 1 suitable hairs can be analysed for the presence of both nu-DNA and mt-DNA. However, nu-DNA in the cortex has been shown to be highly fragmented, present in very small quantities and difficult to extract (Edson *et al*, 2013).

In some hairs small air inclusions, called *cortical fusi*, can be seen between the cortical cells. Often these are seen closer to the root end of the hair. They start out life as fluid filled spaces in the bulb region and become air filled spaces as the hair hardens and dries out. Cortical fusi should not be confused with being pigment granules as careful observation shows them to be between the cortical cells.

In hairs that are coloured the cortical cells are where most pigment will be found. Hair colour and pigmentation are considered separately in Section 2.7.

2.6.2.3 Medulla

Harding and Rogers (1999) describe the development of the **medulla** from the germinative cells of the bulb adjacent to the apex of the dermal papilla resulting in the emergence of electron dense amorphous granules. These granules eventually merge or fuse creating large intercellular spaces with the cell content coalescing at the cell periphery. In mature hairs the medulla forms a central core in the hair shaft which may be interrupted if there has been a pause in the production of medullary cells. The protein in medulla cells is rich in the amino acid citrulline.

In many non-human animal hairs, the medulla has a definitive and regular appearance or pattern that can be described and can be characteristic of the animal species. The apparent structure of the medulla is due to the cells of the medulla collapsing with the medulla appearing as a network of cellular connections with spaces filled with air (Rogers, 1964). Animal medullas will be further considered in Chapter 4 but only in so far as is necessary to clearly identify hairs as being of non-human origin.

In human hairs a medulla may not be present or at least readily visible. In some coarser hairs the medulla may be essentially continuous although often not present at the immediate root end and tip. More often the medulla when present will be discontinuous. The degree of "discontinuity" can vary from short interruptions to the medulla to very small section of visible medulla. The medulla of human hairs occupies no more than one third of the shaft diameter or less.

Human hair medulla is generally unstructured or amorphous in appearance. When viewed with TLM after being placed in a suitable mountant, the medulla appears dark and often opaque—this is due to air being trapped in the medulla. If the medulla becomes filled with mountant the air can be replaced with liquid and the medulla appears translucent or may not be clearly visible. Care needs to be taken when classifying a human hair medulla as opaque or translucent for forensic purposes as its appearance may be more of an artefact than a real feature.

2.7 HAIR COLOUR AND PIGMENTATION

2.7.1 Visual and Macroscopic Assessment of Natural Hair Colour

The assessment of overall hair colour is of interest to scientists in many fields of study including physical anthropologists, cosmetic scientists and medical scientists (Suter, 1979, Hassall *et al*, 2018). The Fischer-Saller Scale (Fischer and Saller, 1928) is used by physical anthropologists to assess visual colour under controlled light conditions. Using this scale hair colour is subdivided into eight main categories these being very light blond, light blond, blond, dark blond, light brown-black, dark brown-black, red and red blond. On a world scale dark brown and black account for about 80% of people.

The perception of the visual colour of bulk hair of an individual human is an overall impression that depends on factors such as the lighting conditions and is an average colour comprising numerous individual hairs. For individual hairs the perceived colour is a product of reflection and refraction of light. As Harding and Rogers (1999) state, a requirement in forensic examinations and comparisons is a well-defined method for mounting individual hairs and their examination under controlled lighting conditions.

With respect to the assessment of colour of hairs in a suitable microscope mountant (this will be discussed in Chapter 4) the key information that the Fischer-Saller scale tells us is that there only are four actual colours these being blond, brown, black and red. The other subdivisions relate simply to shade. In the study of Hassall *et al* (2018) these authors added grey as fifth visual colour. Robertson (1999b) has compared previous studies of the forensic classification of hair colour and from this developed a checklist for assessing colour of mounted hairs. In this scheme hair is classified as colourless, yellow, brown, reddish, black and opaque. In Chapter 4 of this book the authors present a modified version of this classification that also adds grey as a colour category. In the Robertson scheme, the observer is also asked to note shade as light, medium or dark and to assess colour along the length of the hair shaft.

Hair colour is known to vary in the early years (Prokopec *et al*, 2000) but is then generally assumed to remain stable until the age-related onset of greying or whitening, technically called *canities*. The age of onset of greying is variable. Rook and Dawber (1982) report white hair first appeared in Caucasians at the age of 34.2 plus or minus 8.6 years and in African hair on an average 10 years later than in Caucasians. Body area is also a factor with beard and moustache hairs becoming grey earlier than scalp or body hair.

Of course, visually grey hair does not mean that all hairs are devoid of pigmentation. The visual appearance of grey is normally caused by a mixture of pigmented hairs and hairs with either reduced levels of pigment or no pigment.

In conclusion, assessing hair colour of non-mounted hairs in a forensic context can only ever give a very broad classification of perceived colour and should be treated with caution with respect to differentiating hairs. Hair colour should be assessed for individual hairs in a suitable mountant under defined and controlled light conditions.

2.7.2 Pigment Basis for Hair Colour

The colour of human hair is determined by the pigmentation although, as we have seen above, influenced by viewing conditions. At a microscopic level of examination, using LPM and epi-illumination or using TLM, hair colour is principally about the amount of pigment present and the colour of the pigment. Hence, an essential to understanding colour in hairs is to understand the nature of pigments found in hairs.

In hairs pigment granules are produced in the follicle bulb by the Golgi apparatuses of the **melanocytes** located close to the dermal papilla (Tobin *et al*, 2005). The melanocytes release membrane enclosed **melanosomes**, which move into the differentiating cortical cells through a phagocytic process via the melanocyte dendrites. As the hairs are keratinised the melanosomes are embedded in a matrix of keratin associated proteins (Tobin, 2009). Pigment granules are not common in the medulla or cuticle of hairs shafts but are sometimes present. Pigment granules are on average relatively evenly distributed along the hair shaft except near the root end in telogen hairs where pigment synthesis has ceased and sometimes also at the tip end of hairs with a natural taper. Distribution across the hair shaft tends to be towards the outer cuticle.

There are two types of melanin pigment present in hairs these being **eumelanin** and **phaeomelanin**. Both types of melanin are derived from the amino acid tyrosine but with different biosynthetic pathways. Harding and Rogers (1999) provide greater detail on these biosynthetic pathways. Eumelanin results in dark brown to black colours and phaeomelanin in yellow to reddish colours. The overall colour of an individual hair is determined by the relative amounts of these two pigment types. Hence, in brown or black hair eumelanin's predominate and in yellow to reddish hair phaeomelanin's predominate (Jimbow *et al*, 1991). Eumelanin produces pigment granules that are ellipsoidal in shape and about 0.8–1.00 µm long and 0.3–0.4 µm in diameter (Swift, 1977). Phaeomelanin granules are smaller and more spherical in shape (Montagna and Parakkal, 1974).

CHAPTER **3**

Recognition, Recording and Recovery Considerations

3.1 INTRODUCTION

As it has been discussed in Chapter 2, whenever there has been human or animal involvement at the scene of an alleged crime, because of the shedding or loss of hairs, it is almost certain that hairs will be present. The primary responsibility lies with the *crime scene examiner* (CSE) to decide what to collect at the scene. This **evidence recognition** cognitive skill is central to the effective, as well as the efficient, processing of a crime scene. It is simply neither desirable nor practical to collect everything at a scene. Australian Standard AS5388.1 outlines five principles that should guide the collection and subsequent management of physical materials which include hairs. These are as follows:

- Recovery is relevant and optimal.
- The integrity of the physical material is not compromised.
- The potential for contamination is minimised.
- Evidence continuity and security is maintained.
- The potential for analysis is optimised.

Hairs may also be collected from persons and here the standard emphasises that such examinations should be carried out with appropriate consideration for the dignity and well-being of the person involved. There will normally be jurisdictional legal requirements relating to the collection of hairs from persons that may determine who can collect hairs and under what circumstances. In the investigation of sexual offences, the relevant person is most likely to be a medical professional.

The key point is that unless the investigator, the CSE and the medical professional believe there is value in the examination of hairs then they may, by lack of accurate knowledge, fail to recognise the value of hairs and hence be less than committed about making the effort to collect hairs.

DOI: 10.4324/9781315210650-3

The purpose of this chapter is not as an introduction to crime scene examination but to stress that hair examination starts at the scene or with person or persons of interest and that the success or failure of any subsequent examination is critically dependent on the attitude and skills of those involved in making the decision to collect or to not collect hairs.

3.2 RECORDING AND RECOVERY OF HAIRS

3.2.1 Scene Considerations

Regardless of the specifics of a scene the overall approach must be systematic, objective, thorough, planned and documented.

As far as possible all physical material should be recorded *in situ* before being recovered or collected. This should take in to account the condition, position and location of physical materials.

Recording will include taking appropriate notes and still and/or video recording. AS 5388.1 provides guidance on the standards that should be met with respect to recording. The use of sketches and scale plans are also useful in showing

- the location of hairs *in situ* prior to collection,
- the location of hairs relevant to other objects or materials present, and
- the overall physical environment.

The location of hairs and any information as to how they may be present can provide vital information in assessing the relevance of their presence and assist in interpreting how the hairs came to be in the precise location. This type of information may be the most critical aspect of a hair examination in determining what relevance hairs may have to the circumstances of an incident.

The CSE must decide to collect hairs at the scene or to leave hairs on items and to collect the item for subsequent laboratory examination. For example, in examining the scene of an alleged sexual offence should hairs located on bedding be collected at the scene. In the view of the authors it almost always better to collect visible hairs at the first opportunity due to the potential for hairs to move from their original position or to be lost from the item. The same situation applies to hairs on a person. During the removal of a deceased person to a mortuary there is the very strong possibility of hairs being moved and or lost. It is important to again stress that we are not suggesting that a CSE collect every hair at a scene without thought! Indeed, we are suggesting the opposite, that is, the skill of the CSE is to determine what is most likely to have potential evidential value. However, this selection process will only be successful providing the CSE believes that there is value in collecting hairs.

Where it is impractical to collect hairs at the scene every effort must be made to protect them from being lost. This may involve the use of specialist bags to protect hands or feet or other areas.

3.2.2 Scene Sampling Protocols

When sampling for hairs consideration should always be given to minimizing the potential for contamination. AS 5388.1 points out that where there is potential for contamination:

- Victim and suspect samples shall be collected separately (i.e., separated by time and space unless already comingled at the scene).
- Collection equipment shall be discarded after each use. If this is not possible, then collection equipment shall be decontaminated after each use.
- Appropriate personal protective equipment (PPE) shall be worn.
- Collection equipment and packaging should be sterile and free of detectable DNA contamination.

As hairs are a valuable potential source of DNA it is essential that they are collected in such a way as to minimise the potential for contamination through inappropriate handling.

Regarding the collection of hairs at a scene this must be guided by the specific circumstances at the scene and the nature of the items from which hairs may be recovered. However, as general rule, if a hair or hairs are visible then they should be collected by hand—although called "handpicking" hairs should never be picked up with bare hands due to the potential to contaminate the sample. The CSE should also note any special circumstances relating to the interaction of the hair(s) with the surface from which they were collected. For example, were the hairs loose on the surface or were they held in the weave of a garment?

Individual hairs or a small group of hairs collected by hand should be placed in a paper "boat" as shown in Figure 3.1.

The CSE may also choose to recover remaining hairs using tape lifts. These are most likely to be shorter hairs that are less visible or easily removed. It is important that tape lifts are not overloaded as the subsequent laboratory examination of overloaded tape lifts is highly inefficient. Tape lifts must never be stuck directly onto paper!

In general, the use of vacuuming or sweeping to recover hairs is NOT recommended and should only be used to access difficult to get to locations. The use of these techniques is to be discouraged as it destroys

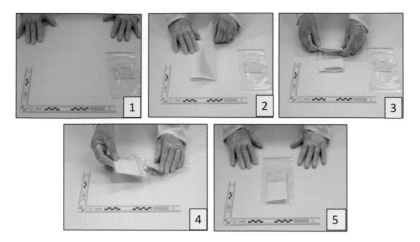

FIGURE 3.1   Construction of a paper boat collection kit.

CONSTRUCTION OF A "PAPER BOAT" COLLECTION KIT

1. Prepare a clean surface using 10% bleach and 70% ethanol. Wearing disposable gloves, place a single sheet of A4 printing paper onto the cleaned surface. Prepare a snap seal plastic bag (SSPB) with an appropriate collection label attached, place beside the sheet of paper.
2. Fold the A4 paper into thirds longitudinally.
3. Fold the already folded paper into thirds again, horizontally.
4. Place double folded paper "boat" into SSPB.
5. Close the seal on the SSPB and a hair collection "paper boat" is prepared.

To collect hairs into this paper boat, unfold until the A4 sheet is fully opened, place the hair sample/samples within the centre of the folded areas, secure individual hairs with Post-Flags™ (see Figure 3.2) and refold the paper along the original fold lines; place the boat back into the SSPB, seal with evidence tape and note location, etc., on the label.

Several individual hairs may be placed on one boat with hairs held under a small Post it Flag™—hairs held in this way are easily removed for examination in the laboratory (see Figure 3.2). The paper boat should then be placed in a plastic snap seal bag. Hairs should never be placed under Sellotape on paper. (Note: this term is used in a generic sense to mean all clear adhesive tapes and does not endorse a brand.) We also do not recommend placing hairs directly into plastic sample bottles—perhaps due to the fact that pathologists routinely use sample bottles, hairs collected at post-mortem (PM) examinations are often in this type of container.

FIGURE 3.2 Securing a hair sample within a paper "boat" collection kit. Diagram represents a hair sample being secured within a paper boat using a Post-it Flag™. This achieves the security of the hair from accidental loss, the means of examining an individual hair using the stereo microscope without touching the hair and adding brief notations if required.

the ability of the hair examiner to consider the location of recovered hairs in interpreting the potential evidential value of recovered hairs. This approach should not be used simply because "there were a lot of hairs to be recovered".

Recently one of us has seen the use of the so-called "Swiffer" pads to recover hairs at a scene. These are commercial floor cleaning wipes and, in the specific case, were used to recover hairs from a wooden floor. The problem with this approach is that only a broad location for any recovered hair is possible—in this case the location of the hair was potentially of great significance. Where such an approach is used only well-defined areas of limited size should be sampled with each pad.

3.2.3 Sampling from Deceased Persons or Remains

The examination of deceased persons or remains may take place at a scene, as part of an exhumation, or part of a PM examination at a mortuary. From the perspective of physical evidence an important consideration in deciding what to collect at the scene will be the potential for loss of physical evidence once the body has been moved. As we have stressed previously the key is the potential loss of accurate information as to the location of physical evidence.

It is normal practice in many jurisdictions for a CSE to attend at PM examinations. The role of the CSE at the PM may be to assist with recording, taking photographs and assisting as required with the collection of potential evidence. The role of the CSE in advising the pathologist is an important role as the CSE may be more knowledgeable as to the requirements of the laboratory for subsequent examinations.

As well as potential **extraneous** physical materials such as hairs, the collection of **known** samples will normally take place during the PM process. Known samples of hairs should be collected where the case circumstances suggest that it would be useful. In our view the default position should be to collect known samples of hair following the GIFT principle (Get It First Time) as if it later becomes a relevant consideration it may be too late to collect known samples. This brings us to the consideration of what is a relevant known sample.

The first guiding principle is to visually assess the deceased person and their hair. Known hair samples should be **representative** of any visual variation. This means that hairs would rarely (if ever) be selected from a single location in the scalp or any other body area.

For known **scalp hair** at least 100 hairs should be removed, selected to represent visual variation in colour and length. These hairs should be *combed* out so that the hairs will include hairs in all growth stages. If hairs are plucked, then anagen hairs will be over-represented. Only in exceptional circumstances should hairs be cut and, if so, NOT from a single location and as close to the skin level as possible. A lock of cut hair may look adequate but is highly unlikely to be representative. If too few hairs are collected this will almost always limit the subsequent microscopic examination process and the conclusions that can be drawn.

Prior to the collection of known **pubic hairs**, the pubis should be gently combed to recover any loose hairs that may include foreign hairs of potential evidential significance. Following this process 40–50 pubic hairs should be recovered by more vigorous combing and, if necessary, plucking.

The recovery of know hairs from other areas of the face or body will usually be determined by the circumstances of alleged incident. However, in our experience, insufficient consideration is often given to the collection of other known hairs.

Once collected, hairs should be placed in a paper boat and NOT in plastic sample bottles, directly into SSPBs or onto small pieces of Sellotape.

3.2.4 Sampling from Living Persons

Sampling of hairs from living persons will usually involve medical examinations of alleged victims and suspects. These procedures will be governed by the relevant jurisdictional specific legal and forensic standards, often incorporated into specific forensic procedures legislation. The later will usually define what is an **intimate** and a **non-intimate** sample or procedure and determine who is authorised to take such samples or conduct such procedures. CSE and other players need to have a detailed knowledge of legislative requirements in their jurisdiction.

With respect to alleged sexual assaults examination and collection of samples, this is usually carried out by a *forensic medical officer* (FMO). As previously stated, regardless of the legislative requirements, ethically these types of examination should consider the dignity and well-being of the person involved. It needs to be recognised that, regardless of the potential legal end point of an investigation, the first responsibility of a medical officer is to their patient. The victim of a sexual assault is already traumatised. The process of a thorough sexual assault medical examination has the potential to add to that trauma, especially when it comes to taking hair samples. An important role for the FMO is to explain procedures to a victim so that they can make a fully informed decision as to what examinations they are willing to consent. An important role for the CSE and/or forensic hair examiner is to provide contemporary information to the FMO so that they can in turn provide accurate information to victims. As previously stated, meaningful laboratory examinations can only be conducted if relevant known hair samples are taken during medical examinations.

Guidelines for known scalp and pubic hairs are as previously described for deceased persons except that for pubic hairs plucking is to be discouraged due to the potential traumatic impact on the individual. If necessary, hairs can be cut at skin level.

All sexual assault examinations must use sealed sexual assault examination kits that comply with ISO 18385:2016 (Minimizing the risk of human DNA contamination in products used to collect, store and analyse biological material for forensic purposes)—requirements. Appendix F of AS 5388.1 provides guidelines for the use of a forensic medical examination kit (FMEK) to collect biological samples for forensic analysis.

Regarding the examination of suspects, the procedures to be followed with respect to hairs do not vary from that described above and should also be carried out using a sealed FMEK.

Appendix 3.1 is an example of our collection guidelines.

Once again, it is worth a reminder that it may be relevant to collect other known hair samples as informed by the circumstances of the case.

Males possess on average more body hair than a female and consideration should always be given to taking samples of body hair. Body hairs are often present in cases of alleged sexual assault but are all too often ignored because of the perception that they are little or no evidential value. Whilst it is true that body hairs possess less microscopic features than scalp hairs, this does not preclude microscopic examination and body hairs can yield DNA profiles.

It is also relevant to recognise that it is now a common practice for persons to partially or wholly remove their pubic and/or body hair and that it may simply not be possible to obtain a known sample in these circumstances. The FMO should note such information that should be available to the forensic hair examiner.

Finally, victims of alleged crimes, and sometimes persons who become suspects, will present at emergency departments of hospitals. They may also have arrived by ambulance with the involvement of paramedics. In these settings the primary, and perhaps sole, focus of paramedics, nurses and doctors will be on the health and welfare of the patient. Whilst the latter is unarguably correct, with a level of appreciation of the forensic aspects that may follow, it is possible to also collect clothing or other items in such a way that potential forensic evidence is not wholly compromised or simply lost. The emergence of *forensic nursing*, especially in emergency departments and in specialist sexual assault centres, should improve procedures to meet this secondary role for these professional groups. It is incumbent on the forensic profession to include these groups in forensic awareness training to improve understanding of how the forensic aspects can be incorporated without compromising medical procedures or ethics.

3.4 CONCLUSIONS

It is a statement of the obvious, but sadly one that needs to be made, that with hair examination it is not a case of "rubbish in rubbish out" (although that also applies), it is all too often a case of "nothing in, nothing out". Unless the value of examining hairs is understood then hairs may simply not be collected. This starts with the investigator, then the CSE, the laboratory scientist and then the legal players. As is often quoted when discussing physical evidence (Harris v United States, 331 U.S.145, 1947), "only human failure to find it, study it and understand it, can diminish its value".

Hence, the focus and message from this short chapter is to stress the vital role of the CSE as the key individual in recognising the presence of hairs as items of potential evidential value and hence then ensuring they are properly recorded and collected/recovered for subsequent examination. The CSE needs to understand that hairs can be a source of DNA (both in animal and human hairs) and that hairs can have a valuable criminalistics role assisting to answer the what happened question. In the latter the CSE plays a vital role in ensuring that due attention is given to the location of recovered hairs and how these hairs may be retained.

Other professionals, especially in the medical profession also need to give appropriate attention to the recovery of physical materials, including hairs that may be of potential evidential value. These professionals will most often be responsible for the collection of known samples that, properly taken, underpin subsequent laboratory examination but not properly taken will undermine or limit such examinations.

In the next chapter we turn our attention to the laboratory examination of hairs.

APPENDIX 3.1: FORENSIC KNOWN HAIR COLLECTION KIT

Collection Instructions

1. Spread out A3 piece of paper on a bench or table.
2. Ask individual to lean over paper so their head/hair is over the middle of the paper.
3. Using a "nit" comb, comb from back to front of the head—see Figure 3.3—Image 1 and Image 2.
4. Repeat Step 3, but this time comb hair from front to back.
5. Between each combing session remove any hair from comb that has not fallen onto the paper.
6. Ensure there are at least 50 scalp hairs—see Image 3.
7. If combing is NOT an option, wearing gloves, carefully cut as close to the scalp as possible collecting a number of hairs from at least five different head regions—see Figure 3.4 and Image 4. Place these inside the unfolded paper from the snap sealed plastic bag.
8. Re-fold the paper and place back into labelled snap sealed plastic bag, seal, sign and date.
9. For body and pubic hairs, comb if possible, or cut as in Step 7.
10. If collecting a known set of body hairs, ensure they are packaged separately and clearly labelled. Common areas for body known sets include: arms, legs, pubic, chest and back.

NOTE: Please DO NOT shave any areas—comb or cut.

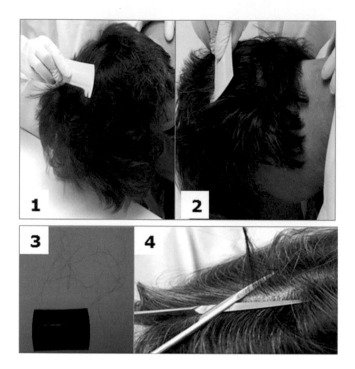

FIGURE 3.3 Back combing and cutting hair.

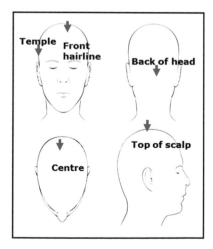

FIGURE 3.4 Scalp regions for collecting cut hair samples.

Collection Kit Includes:

2× pieces A3-sized paper
2× sets latex gloves
2× snap sealed plastic bags with labels
1× fine toothed "nit" comb
1× pair of scissors
1× instruction sheet

CHAPTER 4

Laboratory Examinations

4.1 INTRODUCTION

It is assumed that the reader is familiar with the broad quality requirements to meet ISO/IEC 17025:2018 (general requirements for the competence of testing and calibration laboratories). This standard replaced the previous edition (ISO/IEC 17025:2005) with the main changes being a move to a risk-based approach and greater flexibility in the requirements for processes, procedures, documented information and organisational responsibilities. In this chapter, reference will also be made to AS 5388.2-2012 (Forensic Analysis. Part 2: Analysis and examination of materials).

In the contemporary environment, the approach taken by organisations and individuals to the examination of hairs will vary depending on the scope as defined in the laboratories quality system. All laboratories will have some form of management decision-making or a triage process that will define the purpose and the extent to which the laboratory will examine hairs. The scope for examinations may range from not conducting any examination of hairs, selecting what are thought to be human hairs for DNA testing, some form of preliminary examination or triage to separate and/or select human hairs from non-human (hereafter, animal) hairs, low-power microscopic examination (LPM) to eliminate hairs from further examination, to detailed microscopic examination. Beyond this level of examination there may be deep subject experts who have specialist knowledge not usually applied in forensic examinations.

In this chapter, we will look at protocols and practice dealing with what we term level 1, level 2, level 3 and level 4 examinations as follows:

Level 1: Recognition and separation of human and animal hairs.
Level 2: Examination of human hairs including body area and ethnic origin and selection of hairs for DNA analysis.
Level 3: Detailed examination of hairs and comparison microscopy.
Level 4: Specialist examinations.

DOI: 10.4324/9781315210650-4

The starting point for this chapter assumes that hairs have been recovered/collected at the scene of an alleged crime, or items submitted to the forensic laboratory for examination that may include hairs as a physical material, including reference or known samples of hair. Such samples should meet the required standards for acceptance by the laboratory.

AS 5388.2 details criteria for the acceptance of physical material received for examination. Material may be rejected on the basis that

- packaging is inappropriate such that the integrity of contents is compromised,
- it is inappropriate for the testing requested—this could include the lack of known samples,
- the examinations requested are not relevant to the investigation,
- continuity has been compromised, and
- the analysis is unlikely to yield evidence of probative value or **less likely to provide evidence of probative value than other evidence submissions** (our emphasis).

A commonly held perception is that the examination of hairs is time consuming relative to the outcomes. However, in our view, provided the laboratory has a well-thought-out case management/triage system to accept or reject submissions that results in only material relevant to each case being accepted and that any such materials are examined in a logical sequence, then meaningless examinations of hairs should be avoided maximising the potential for useful outcomes.

Following the advice in AS 5388.2, which states that "the facility should make appropriate scientific and/or technical personnel available to discuss analytical needs and likely outcomes with clients" and further recommend that "conferences involving representatives from all interested parties may be appropriate to discuss what analysis and/or examinations are to be carried out and in what order", should result in a well-focused use of resources and a shared understanding by all parties of the potential evidential value of such examinations.

In our view, many of the perceived problems with hair examination come from unrealistic expectations and subsequent disappointment or frustration with the outcome. We believe firmly that there is probative value to be derived from the examination of hairs, but that to realise this value all parties need to be clear as to the purpose of examinations and any limitations inherent in the examination process. Quality is about getting the right result in the context of it being **fit for purpose**—that purpose needs to be agreed.

Our level 4 protocol provides a framework for defining the purpose of examinations and reducing the potential for unrealistic expectations and perceptions of failure. An understanding of scope and limitations is also key to defining error rates—we will return to this topic in Chapter 5 when dealing with interpretation and reporting.

4.2 LEVEL 1 EXAMINATIONS—RECOGNITION AND SEPARATION OF HUMAN AND ANIMAL HAIRS

It is quite common that in cases involving requests for hair examination, the first question is to determine that the "fibre" is indeed a hair and then whether it is of human origin. This may involve looking at a single hair to looking at very large numbers of hairs. Although the scope of this book is largely limited to the examination of human hairs, the examiner must have enough knowledge of animal hairs to be able to differentiate human from animal hairs—this is a basic requirement and skill.

Later in this chapter, we will consider variation in human hairs related to body area and ethnic origin.

For animal hairs, different types of hair can be present in the fur or pelage. Usually these are visually clearly different based on the degree of coarseness with the most common types being coarser **guard-hairs** and finer **under-hairs**. Guard-hairs are usually longer and generally display the widest range of microscopic features which makes them the most useful for identification.

Initial examination of questioned hairs should start with a visual examination of the hairs and then examination of unmounted hairs using LPM. Whilst we make no recommendation as to the make and model of microscope, we use a Leica stereo microscope with a ring light source to ensure even illumination, as illustrated in Figure 4.1.

Based on the **shaft profile**, it is often possible to quickly separate human and animal hairs. The shaft profile of human scalp hair is generally quite uniform with only moderate variation in shaft diameter along the length of the hair shaft. Whilst non-scalp hairs from humans do show greater variation in diameter and shaft profile, they do not display the typical shaft profiles seen in animal hairs. Notwithstanding, care needs to be taken when examining a sample of hairs of mixed animal and human origin especially if non-scalp hairs or hairs from non-Caucasians are present. Some straight animal hairs may also not be readily distinguished on visual examination.

Appendix 4.1a shows a proforma checklist that can be used to record the visual and LPM features for animal hairs. We strongly recommend the use of checklists to ensure a systematic, thorough and appropriately detailed approach to the examination and recording of hair features. The shape of the shaft of non-human hairs can be classified as **shield, straight, symmetrically thickened** or **wavy**. Examples of these are shown in Figure 4.2.

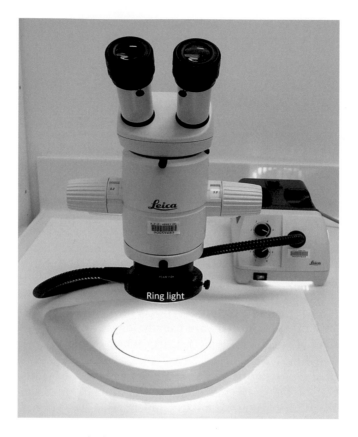

FIGURE 4.1   Stereo light microscope/low-power microscopy (LPM).

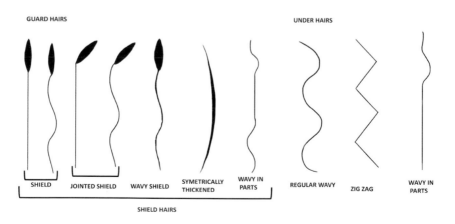

FIGURE 4.2   Diagram of macroscopic animal hair profiles.

The only other feature recorded at this stage is overall **colour**. In many animal hairs there will be profound changes in colour along the hair shaft called **banding** (see Figure 4.3 for examples of animal hair banding). In naturally coloured human hairs such banding will not be seen.

In Chapter 2, the basic structure of hairs was described with three anatomical components, these being the outer *cuticle*, the main body of the hair shaft the *cortex* and, variably present, a central *medulla*. Differentiating human and animal hair requires an understanding of the microscopic appearance and variability of these three components as seen in human and animal hairs.

Table 4.1 summarises the main points of difference that are the basis for differentiating human and animal hairs. Most of these features

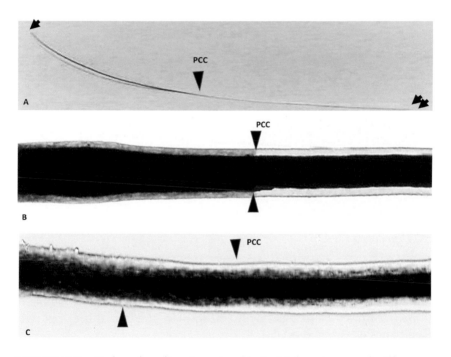

FIGURE 4.3   Colour banding in animal hair. Light micrographs illustrating the profound colour changes (PCC) that occur naturally in animal hairs. (a) The hair here is from an Alsatian adult female and is photographed on white paper as visualised. The single arrowhead is the root end and the double arrowhead is the tip end of the hair. The larger arrowhead indicates where the PCC occurs. (b) This light micrograph is the same hair as A where the PCC (arrowheads) is more pronounced. (c) Also a hair from the same dog but the colour change (PCC) is less defined, tending to blur into the reddish brown from the white areas (arrowheads).

TABLE 4.1 Comparison Features of Human and Animal Hairs

Feature	Human Hairs	Animal Hairs
Hair type	Scalp hairs Body hairs	Guard hairs Under hairs—wool or fur hairs Tactile hairs—whiskers Other—mane, tail hairs
Root shape	Three types only—indicates growth stage: Anagen Catagen Telogen	Highly variable—often indicative of animal family
Colour and Pigmentation	Generally consistent throughout shaft	Hair shaft banded, colour and pigmentation may exhibit radical colour changes in a short distance (banding)
Distribution and density of pigment	Either evenly distributed along shaft or slightly denser towards the cuticle	Centrally distributed, generally denser towards the medulla
Medulla	When present—amorphous in appearance, generally less than one third of overall shaft diameter	Normally continuous and structured, generally occupies greater than one third of shaft diameter
Hair shaft profile	Straight or curved Wavy Curly Peppercorn	Shield Straight Symmetrically thick Wavy
Cuticle	Scale pattern routinely imbricate	Variable cuticle scale patterns, hair shaft more variable

require that the questioned hair is mounted for microscopic examination although some features can be seen if the hair is examined unmounted at the LPM level. Here it can be useful to place the hair(s) between two glass plates or two glass slides.

The usual method to examine scale patterns in animal hairs is to make a scale caste. For the limited purpose of eliminating animal hairs this can usually be achieved without considering the scale pattern features.

In selecting a suitable mountant for hair examination, the major consideration should be refractive index (RI) with this being in the range of 1.50–1.60 (Roe *et al*, 1991). The authors use Hystomount™ with a RI of 1.55. Hystomount™ is a permanent mountant that will dry out over time to a hard-solid material suitable for the long-term storage of slides. An alternative is to use glycerine/water (50:50 v/v) as a temporary mountant as this allows hairs to be easily recovered should a scale caste be necessary. Water is not a suitable mountant for the examination of human hairs as it has a RI which is too low to enable visualisation of internal detail.

Each recovered/questioned hair and known hair should be individually mounted on separate slides. Where hairs are short a small number of hairs, usually no more than five, may be placed on a single slide with suitable identification. Placing a large number of known hairs on a single slide is to be discouraged.

Based on the features summarised in Table 4.1, it is relatively easy to separate animal hairs from human hairs. In a screening of numerous co-mixed animal and human hairs any animal hairs that are not readily excluded at a visual or LPM level of examination can be mounted in a temporary mountant and examined at LPM—in most situations any remaining animal hairs will be differentiated at this level of examination.

The detailed examination of animal hairs is out of scope for this book. However, the features that can be assessed with a more detailed examination are shown in checklists Appendix 4.2a and 4.2b. Appendix 4.3 (Robertson, 1999) describes these features. One of the features which is readily seen with LPM, and higher-power transmitted light microscopy (TLM), is the medulla of animal hairs. As indicated in Table 4.1, the medulla can vary along the length of an animal hair shaft. This is commonly seen in hairs with a shield profile, where the narrower shaft at the root end may have a single seriate ladder medulla that becomes a multiseriate ladder medulla as the hair shaft broadens. Figure 4.4 shows some examples of the two major types of medulla seen in animal hairs, these being a *lattice* medulla and a *ladder* medulla.

The examination of animal hairs is a specialist aspect of hair examination that falls into the category of a level 4 examination. Whilst there is an extensive literature dealing with the microscopic identification of animal hairs, few studies have a specific forensic focus. Hence, most studies are limited to including hairs relevant to the context of the study and often considers only those microscopic features that are useful to identify hairs in that context. Typical studies include food contamination focusing on identifying rat or mouse hairs, or wildlife studies focused on the local population of interest. Citing only two references as examples, Teerink (2003) has published an atlas and identification key for hair of West-European mammals and Brunner and Triggs (2002) an interactive tool for identifying Australian mammalian species called "hair ID".

Even for the limited scope of eliminating animal hairs, it is strongly suggested that the laboratory build up a *reference collection* of common animal species. This could include domestic pets such as cats and dogs, local farm animals and other species relevant to the geographic location or specific local factors. The hair examiner should at least be familiar with the macroscopic appearance of commonly encountered hairs and be able to effectively triage hairs that may require more detailed examination by a level 4 expert. The laboratory should identify local experts who may be able to assist if required. Ideally a relationship should be established such that the external provider has an appropriate understanding of what is required to meet forensic standards. For example, ISO/IEC 17025:2018 Section 6.6 requires that "the laboratory shall ensure

MULTI SERIAL   WIDE MEDULLA   WIDE AERIFORM   UNISERIAL
LADDER   LATTICE   LATTICE   LADDER

FIGURE 4.4   Animal hair medulla types. Light micrographs illustrate the more common types of medullas observed in animal hairs. They include the *multiserial ladder*; the *wide medulla lattice*; the *wide aeriform lattice* and the *uniserial ladder*. Each type of medulla may be either wide or narrow depending on the animal species except the uniserial ladder which is generally only found in animal under hairs.

that only suitable externally provided products and services that affect laboratory activities are used". In Section 6.6.3 the standard requires laboratories to communicate its requirements to external providers for

   a. the products and services to be provided,
   b. the acceptance criteria,
   c. competence, including any required qualifications of personnel, and
   d. activities that the laboratory, or its customer, intends to perform at the external provider's premises.

Descriptions of some common animal hairs can be found in Robertson (1999b).

Finally, not only at level 1 but even at level 2 and level 3, unless an examiner has received specific training in the identification of animal hairs, caution should be exercised, and a conservative approach adopted even for the limited purpose of elimination of animal hairs from further examination.

Hair examiners should be aware that there are commercial services available to conduct both nuclear DNA (nu-DNA) and mitochondrial DNA (mt-DNA) on animal hairs and be able to advise investigators as to the current services available should this testing be required.

4.3 LEVEL 2 EXAMINATIONS—EXAMINATION OF HUMAN HAIRS FOR BODY AREA AND ETHNIC ORIGIN AND SELECTION FOR DNA ANALYSIS

The initial examination of human hairs includes a visual assessment of the hair(s) and the use of LPM of unmounted hairs. At this level, the examiner can make a preliminary assessment of body area determination, possibly ethnic origin, overall visual hair colour, root type/growth phase and the condition of the hair(s), including any obvious disease conditions. These observations should be recorded in a systematic way using a checklist and where necessary written observations. Examiners should be aware of the limitations in what they can see with hairs that are not mounted and examined at the LPM level.

4.3.1 Body Area Determination

Hicks (1977) has listed the features that can assist in body area determinations. These are described in Table 4.2. Note that this publication was updated in 2004 by Deedrick and Koch (2004). These general descriptions need to be interpreted against variation seen in hairs of individuals from different ethnic backgrounds. For example, scalp hair from a person of African origin will not be naturally long as most often the shaft profile will be peppercorn. Scalp hair from a person of Asian origin will not have a soft texture. Furthermore, length and even appearance can be altered in a non-inherited (cosmetic) fashion that can modify shaft profile, colour and the appearance of the tip end. Cruz *et al* (2016), in reviewing the impact of cosmetic procedures and shape-modulating cosmetics on hairs, discussed the factors that influence the shape of the hair shaft. The consensus view would appear to be that shape is determined by the shape of the hair follicle and the distribution and heterogeneity of intermediate and paracortical cells. Wortmann *et al* (2020) support the view that the structural differences of the cell

TABLE 4.2   General Features of Human Hairs from Different Body Areas

Body Area	General Feature
Scalp hair	Long with moderate shaft diameter variation Medulla absent to continuous and relatively narrow when compared to hairs from other body areas Often with cut or split tips, may show artificial treatment, solar bleaching or mechanical damage such as caused by backcombing Relatively soft texture (pliable)
Pubic hair	Shaft diameter coarse with wide variation and buckling Medulla relatively broad and usually continuous when present Root frequently with follicular tag Tip usually rounded or abraded Stiff texture or wiry
Limb hairs (arm or leg)	Shaft diameter fine with little variation Arc like shaft profile Medulla broad, discontinuous and granular in appearance Soft texture
Beard or moustache hair	Shaft diameter very coarse with an irregular or triangular cross-sectional shape Medulla very broad and continuous
Chest hairs	Shaft diameter moderate and variable Tip long, fine and arc like Stiff texture
Auxiliary or underarm hairs	Resemble pubic hairs in general appearance Shaft diameter moderate and variable with less buckling than pubic hairs Tips long and fine Frequently with bleached appearance
Other	Eyebrow hairs are sabre-like (fusiform) in appearance Eyelash hairs are short, stubby with little shaft diameter variation and sabre-like Trunk hairs are transitional between pubic and limb hairs in appearance

types together with their lateral segregation are the main factors in curl formation.

A relatively recent trend has been for females (although not restricted to females) to reduce or entirely remove body hair and pubic hair (Rowen *et al*, 2016). When pubic hairs are present, and have been cut, their tip end will not have the typical appearance described above.

The descriptions in Table 4.2 provide general guidance to the examiner but still require careful interpretation depending on the case circumstances.

4.3.2 Ethnic Origin

In describing the general features of "racial" subgroups, Hicks (1977) used the terms Caucasian, Negroid and Mongoloid as racial subgroups.

The most recent statement by the American Association of physical Anthropologists (Fuentes *et al*, 2019) states that race does not provide an accurate representation of human biological variation, and that humans are not divided biologically into distinct continental types or racial clusters. As Cunha and Ubelaker (2020) point out "whilst the scientific basis of the original groupings gradually crumbled, the terminology persisted ... the old racial concept of groups being static, pure and fixed gradually gave way to more dynamic, realistic views that recognised processes of gene flow and genetic variation within all groups and areas". These authors state that the recommended terms relating to the three main geographic groups are African, European and Asian.

A study by Koch *et al* (2020) reported that whilst report language is changing from race to ancestry, the use of outdated racial designation is still prevalent amongst hair examiners in the USA. They found that overall accuracy of examiner ancestry assessments was very poor for persons identifying as Asian with over 37% of hairs being incorrectly classified. Accuracy for European ancestry was better with less than 5% misclassified and only 1.4% of hairs self-identified as African were misclassified. However, to further complicate matters, these authors also looked at a sample of hairs from individuals identifying as Hispanic. They concluded that the current framework for assessing ancestry to hair samples from people who self-identify as Hispanic does not align well with genetic or self-identified ancestry.

Hence, we have replaced the older terminology in Table 4.3 with European, African and Asian whilst recognising that the relationship between geographic area and ethnicity is at best a loose relationship.

These are listed in Table 4.3.

The ability to differentiate hairs or assign them to a specific ethnic group is questionable, especially in an increasingly multicultural society. The value in attempting to determine ethnic origin can only be truly assessed against the specific circumstances of a case. In some instances, it may be relatively easy to determine, and the potential value could be significant. For example, if an African scalp hair was found in a case in Australia, this would still be a relatively unusual finding and could be useful in narrowing down suspects. In the United States of America, the finding of an African hair would overall be less significant.

In summary, the hair examiner should be capable of broadly determining body area or somatic origin of hairs and in so doing be aware of the differences between hairs from persons of different ethnic origin.

TABLE 4.3   General Features of Hairs from People of Different Ethnic Origins

Ethnicity	Feature	Variation
European	Hair shaft	Diameter moderate with minimal variation—mean diameter for scalp hairs is 80 μm
	Pigmentation	Pigment granules sparse to moderately dense with even distribution
	Cross-sectional shape	Oval cross section
African	Hair shaft	Shaft diameter moderate to fine with considerable variation. Shaft with prominent twist and curl
	Pigmentation	Pigment granules densely distributed and arranged in prominent clumps. Hairs can appear to be opaque
	Cross-sectional shape	Flattened cross-sectional shape
Asian	Hair shaft	Shaft diameter coarse with little or no variation
	Pigmentation	Pigment granules densely distributed and often arranged in large clumps or streaks
	Medulla	Prominent medulla, broad and continuous
	Cuticle	Thick cuticle

The examiner needs to be cognisant of alterations to hairs that are non-inherited or acquired characteristics and to the fact that we live in a multicultural world where ethnic origin is a somewhat outdated concept. Whilst it will be relatively easy to determine body area, in some instances it will be acceptable to report an inconclusive result or make a non-specific finding such as body hair.

A record of all observations must be made, and this should be systematic and with enough detail for the intended purpose. Essentially all that we are trying to capture and describe is observable phenotypic variation. This record could include written descriptions, sketches or drawings and images. Again, we strongly recommend the use of a checklist to ensure a complete description is taken of all hairs examined. However, it should be stressed that a checklist alone is unlikely to fully capture all the potential information from a thorough examination.

4.3.3 Selection of Hairs for DNA Analysis

This book will not cover the analysis of DNA beyond consideration of specific aspects of hairs as a biological source for DNA analysis. In this

CASE STUDY

The following case study is included to illustrate the importance of body area determinations and also why it is important to just examine hairs!

This case relates to a murder of a male in a shooting at a night-club in Toronto, Canada, in July of 2002. Two men were involved. One was confidently identified as the shooter. This individual was seen leaving the club in a car registered to the mother of the second suspect. Eyewitness identification of this second suspect was not definitive, and the prosecution relied on circumstantial evidence that was said to link this suspect to the murder. Included in the latter were hair clippings from a newspaper in the garbage of the bathroom nearest to the suspects bedroom and hair clippings in an electric razor found in his nightstand. Eyewitness testimony was that the second offender had dreadlocks that were two inches or longer. When arrested the second suspect had very short hair. The prosecution case was that the suspect had shaved off his scalp hair to change his appearance and that this explained why the eye-witness had failed to identify him. The prosecution also argued that the suspect had made an after-the-fact attempt to change his appearance to cover up his involvement in the shooting.

It may seem somewhat surprising, but the fact is that neither the prosecution nor the defence had the hairs examined prior to the trial at which both defendants were found guilty. One ground for the appeal on behalf of suspect 2 was fresh evidence based on examination of the two hairs samples that at his trial counsel "was unaware that it was feasible to perform forensic testing to deter-mine whether hair clippings were from an individual's scalp or another part of the body".

In preparation for the appeal, the hairs were finally examined by two experts for the defence and two for the prosecution.

One defence expert concluded that "it can now be stated with reasonable scientific certainty that the hairs from the clipper and from the newspaper are populations of facial (beard) hair. Neither sample contains a significant number of scalp (head) hairs".

One of the prosecution experts concluded that "there was, as a scientific matter, no evidence to support the proposition that the hair clippings represented a head shave".

All four experts agreed that the hairs from the newspapers were almost all facial hairs. This a clear case where body area deter-mination was neither difficult nor controversial. Whilst we have

no knowledge of the reasoning as to why the prosecution did not
have the hairs tested before trial, it would appear that the defence
council lacked sufficient knowledge of what can be achieved with
hair examinations. The appeal was allowed. We are unaware of the
outcomes a second trial.

This case serves to illustrate why hairs should be examined, and
that body area determination can be of real evidential significance.

(Anon, 2013)

sense, hairs should just be considered as a very useful potential biological
source for DNA that can then be analysed.

In theory, all human hairs are suitable sources for the analysis of
mitochondrial DNA (mt-DNA). Enough mt-DNA can be extracted
from hair shafts as short as 2 mm or less. Brandhagen *et al* (2018) have
shown that levels of mt-DNA decrease along the hair shaft from the
root to tip end. Melton *et al* (2012a) report success rates of 90% or better for naturally shed hairs under 3.5 mm, including hairs that did not
have a root. However, it is highly desirable that hairs are subjected to
microscopic examination before they are submitted for mt-DNA testing
due to the cost of mt-DNA analysis and the limited capacity available
for such testing. It simply makes no sense to submit hairs for analysis
that could have been eliminated as being of interest based on microscopic examination. Hence, in our view, all hairs should be examined
at LPM, preferably mounted, and those hairs that are clearly different
to known hairs eliminated. We will discuss the decision-making process
in more detail when discussing type 1 and type 2 errors. We support
the view of Houck and Budowle (2002) that microscopic examination
and mt-DNA testing are complementary techniques in the examination process. This view is also recommended by Koch *et al* (2020) who
advocate for a combined analytical approach using both microscopical
analysis and mt-DNA data.

The key feature that needs to be assessed in selecting hairs that
may be suitable for nu-DNA testing is the growth stage of the hair root.
Figure 2.3 showed examples of hairs in the three recognised growth
phases, anagen, catagen and telogen.

With the current sensitivity of nu-DNA analysis all hairs that are in
the anagen or catagen growth phase are almost certain to have enough
DNA in their roots to produce a full DNA profile.

In the absence of a root, it is highly unlikely that enough DNA can
be extracted from the hair shaft to obtain a reportable nu-DNA result.

CASE STUDY

A hair recovered from the rear of a vehicle was submitted for examination. The vehicle was suspected of having been used to move the body of a victim of a homicide to a secluded location before the vehicle was dumped elsewhere and when found initially thought to be a stolen vehicle.

Six individuals were identified by police as possible sources for the recovered hair. Four of these individuals were brothers. Microscopic examination of the recovered hair and known samples for three of the brothers, the deceased and two suspects established that all but the deceased could be eliminated on the basis of microscopic examination. Furthermore, the recovered hair displayed features of post-mortem banding contributing information to the context of the case in that the hair could not have been deposited as a result of the victim simply riding in the vehicle whilst he was a live.

Had the recovered hair been sent for mt-DNA testing before microscopy this would have not separated the four brothers who would have had the same mt-DNA and with no indication as to when the hair could have been deposited. This case demonstrates the value of microscopic examination and the complimentary nature of microscopy and mt-DNA testing.

That is not to say that all hairs in the anagen and catagen growth phases SHOULD be submitted for nu-DNA testing! We still strongly recommend that all hairs are examined at LPM, even if not mounted. If a known hair sample is available for comparison, then obviously different hairs should be eliminated. However, we recognise that in many laboratories there is very little, if any, triage process.

In practice, the examiner only needs to be able to accurately recognise telogen hairs. Telogen hairs are not suitable for **routine** nu-DNA testing. This is due to the very low levels of nu-DNA present as well as the usually degraded nature of remaining nu-DNA. However, in some telogen hairs it is possible to extract enough nu-DNA to obtain a full profile or, at least, enough loci for a reportable profile. An approach used to identify hairs that may have enough DNA to produce a reportable nu-DNA profile is to use a suitable DNA stain that makes nuclei visible at a microscopic level of examination. Brooks *et al* (2010) describe a rapid staining technique to detect nuclei using haematoxylin and report DNA success where there are around 30–50 nuclei present. Other researchers have used a fluorescent stain (Bourguignon *et al*, 2008) but this has the disadvantage of taking a longer time and requiring a fluorescence microscope. Lee *et al* (2017) report the use of methyl green. In this study, the authors obtained full or "high partial" profiles from as low as 1–10 nuclei for scalp hair suggesting that it is worthwhile submitting hairs with a lower number of visible nuclei for nu-DNA testing than indicated from previous studies. Interestingly in their study they obtained 100% full or high partial profiles from pubic hairs with more than 50 nuclei. It should be noted that 113 of the 208 scalp hairs they tested had no detectable nuclei as did 42 of the 78 pubic hairs tested. Finally, Ottens *et al* (2013) have shown that direct amplification may further increase the potential to obtain a reportable nu-DNA result.

In summary, with the current approaches to nu-DNA analysis it may be possible to obtain a reportable nu-DNA profile from about 30–40% of telogen hairs. Laboratories should consider incorporating a technique to detect nuclei associated with hair roots as part of their protocol for hair examination. The term "associate with" is intentional as all or some of the nuclei may not be in the hair root but rather in cellular material associated with the hair root.

A study by Brandhagen *et al* (2018) holds out the future promise that it may be possible to obtain nu-DNA results from telogen hairs without roots using high throughput shotgun sequencing and preferential extraction aimed at recovering highly degraded nu-DNA and targeting single nucleotide polymorphisms or SNPs. This study further emphasised that there is nu-DNA present in the shafts of telogen hairs. The challenge is that nu-DNA is degraded to a point where the DNA fragments are simply too short for standard STR-based analysis (Edson *et al*, 2013).

In Appendix 4.4, we present protocols for testing for nuclei based on the study of Brooks *et al* (2010).

4.3.4 Low-Power Microscopic Examination of Hairs

Examination of hairs using LPM can be of mounted or unmounted hairs. Where possible we strongly recommend that hairs are mounted for examination at the LPM level. For unmounted hairs it can be useful to place the hair(s) between glass plates or microscope slides before examination.

The examiner should record the length, shaft profile, colour and the appearance of the root (when present) and tip end of each hair. Variation in features along the length of the hair shaft should be recorded.

The checklist we use for visual and/or LPM examinations of human hairs can be found in Appendix 4.2a.

Hair length is usually of limited value. Notwithstanding, the length of hairs should be recorded by gently stretched or pulling the hair as straight as possible and its length measured. The hair should then be allowed to regain its original profile before the shaft profile is assessed.

When comparing recovered hairs with known hair where there is a major difference in length between a recovered hair and a known hair, the examiner needs to be able to explain this difference if the known hair is NOT to be excluded as a possible source. For example, if the known hair sample is from a female with uniformly very long scalp hair, and the recovered hair is a complete hair and is 5 cm long, then "length" is a useful exclusionary feature.

In drawing any conclusion based on hair length, a key factor to consider is the time gap between the commission of an alleged crime and known samples being recovered. If the time gap is short, hours or days, and the known individual has not altered their hair, then there should not be any significant difference in hair length. If the time gap is significant, weeks to years, then there may be significant differences due to the passage of time that would include hair length.

Any comparison between recovered and known samples is also critically dependent on the known sample being representative of the known individual.

Shaft profile should be assessed of the hair in its natural "relaxed" state. Many scalp hairs will be classified as straight or wavy. There is a degree of interpretation as to what is meant by the term "straight" as few hairs are entirely straight. We include in this category hairs with a slight curve but short of being "wavy". As previously mentioned, "peppercorn" is most often associated with African scalp hair. "Curly" hair can be seen in hairs from the scalp and many body hairs. As we have seen with respect to body area determination eyebrow and eyelash hairs have a sabre-like appearance where the shaft diameter varies to give the appearance of a "sabre" (see Figure 2.6).

Examples of hair profiles are shown in Figure 4.5.

Colour for unmounted hairs is limited to an assessment of visual colour. Visual colour of individual hairs does not always readily equate to colour as seen at LPM of a mounted hair. In terms of the potential to discriminate hairs at the LPM level of examination, colour is the most useful feature assessed. Hence, one reason why it is strongly recommended that hairs are mounted for examination if discrimination based on colour is to be achieved. The foundations for hair colour are discussed in Chapter 2 and are not repeated here. Colour in our scheme is classified as *colourless* (essentially no colour and usually very little pigment present), *yellow, brown, reddish, grey/black* and *opaque* (where the hair is dark due to very heavy pigmentation). As colour may vary along the length of the hair shaft, the colour should be assessed from the root end to the tip end of the shaft. In addition to the colour the intensity of depth/shade should be recorded as light (L), medium (M) or

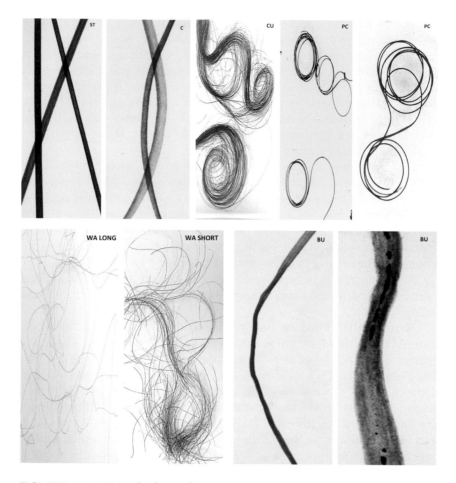

FIGURE 4.5   Hair shaft profiles. Image set A. Light micrographs show the various profiles of the hair shaft observed during hair examination either by eye or LPM. Straight (S) profile or slightly curved (C) shaft profiles; curly (CU) can be open types of curls or quite tight rings, whereas peppercorn (PC) hair shafts are very tightly curled and are so called because they look like black whole peppercorns. Image set B. Light micrographs show the various profiles of the hair shaft observed during hair examination either by eye or LPM. Wavy (WA) hair is neither curly nor curved and buckled (BU) hair has flattened areas along the hair shaft that allow it to and twist or lie in flattened curls.

dark (D) (see Figure 4.6, in six parts). The presence of any artificial colouration should also be noted. The appearance of artificial colouration on the hair shaft will depend on the nature of cosmetic treatment. If the hair shaft is either naturally colourless or the pigment has been reduced or removed by bleaching, then there will be a clear demarcation

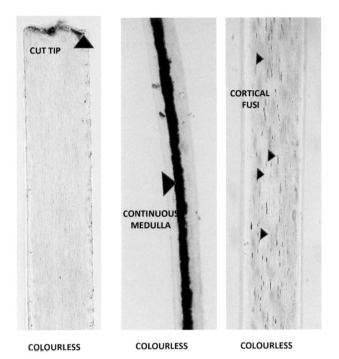

CUT TIP

CONTINUOUS
MEDULLA

CORTICAL
FUSI

COLOURLESS   COLOURLESS   COLOURLESS

FIGURE 4.6 (a) Colourless hair aka white hair. The three light micrographs illustrate microscopically observed colourless or white hair. The colourless state occurs over time as the pigment or melanocytes cease to be produced and laid down in the cortical cells of the hair shaft. Generally, this occurs at an older age-related stage of hair growth cycles but can also occur in much younger persons. The images also indicate some of the features encountered during observation of colourless hairs. The cut hair (arrowhead) example displays a hair with thin cuticle and almost completely lacking pigmentation. The next hair has a dark, continuous medulla (arrowhead) that is in stark contrast to the colourless hair and the third hair example, similarly colourless but displaying many cortical fusi (arrowheads)—also a common feature seen in white hair. *Please note: magnification bars have not been included with all TLM images as the pertinent features have been enlarged for illustration of the feature described. Objective magnification ranges from x20 to x40 for the more detailed images. (Continued)*

between the dyed and undyed portion of the hair (see Figure 4.7). If natural hair pigmentation is visible this may be because the hair has not been subject to heavy bleaching or the dye may be a colour rinse or a semi-permanent dye.

The light micrographs in Figure 4.7 are four examples of artificially coloured hair aka dyed hair. Changing hair colour is almost a "rite of passage" for both men and women. If the hair has been bleached to

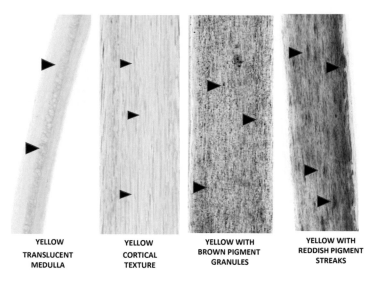

| YELLOW TRANSLUCENT MEDULLA | YELLOW CORTICAL TEXTURE | YELLOW WITH BROWN PIGMENT GRANULES | YELLOW WITH REDDISH PIGMENT STREAKS |

FIGURE 4.6 (*Continued*)   (b) Yellow aka blond hair. The light micrographs here show four examples of microscopically observed yellow-coloured hairs. The first bright yellow hair has a faint translucent continuous medulla (arrowheads); the next yellow hair is less bright but features coarse cortical texture (arrowheads); the third yellow hair displays prominent brown pigment granules (arrowheads); and finally, the fourth yellow hair has distinctive reddish pigment streaks (arrowheads) throughout the cortex. The hairs with brown and reddish pigment granules would possibly be described as "yellow/brown" and "reddish/yellow". (*Continued*)

strip the natural-coloured pigmentation from the hair shaft, a quick test can determine a bleaching episode if required (see Appendix 4.5).

Root type has been previously discussed. The presence or absence of cellular material arising from the sheath on anagen hairs should also be noted. In some cases where hair has been removed with some force it can be useful to count the number of anagen, catagen and telogen hairs and the number of anagen hairs with attached sheath material. Even when hairs have been removed with force not all anagen hairs will have attached sheath material as this is dependent on the individual and the speed with which the hair is removed (King *et al*, 1982). Nonetheless, in a large sample of recovered hairs the presence of sheath cells on most hairs is an indicator of these having been removed with some degree of force.

Another potentially useful feature of hair roots that may be encountered is in hairs from a decomposing deceased person. In a

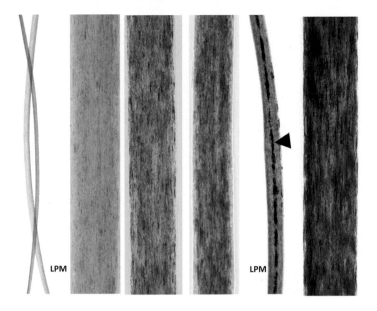

FIGURE 4.6 (*Continued*) (c) Brown hair. The six light micrographs shown here make a lie of the claim that "all brown hairs are the same". In fact, even though brown hair is the second most common visual hair colour it contains probably more accessible (viewable) features/characteristics than all the other hair colours together. The six images here represent only a small glimpse of the variety of brown hairs seen at LPM and at TLM the features observed are many and varied. In one hair at LPM an arrowhead indicates an almost continuous medulla. (*Continued*)

study by Linch and Prahlow (2001), four types of modified root end were described. These were root banding (distal), root banding (proximal), hard keratin points and brush-like cortical fibrils. Examples of the appearance of hair roots from decomposing bodies are given in Figure 4.8. These types of modified root ends were only seen in anagen or transitional hairs and not in telogen hairs with the authors proposing that this may be due to the depth of the hair root in the dermis. The timing of the appearance of the four types was variable, but the earliest onset in their study was after two days—this appeared to be dependent on environmental factors. They concluded that post-mortem changes to the root end are a good indicator of decomposition, but the timing of their appearance does not appear to be helpful in determining post-mortem interval. More recently Hietpas *et al* (2016) have shown that root banding may be the result of the degradation of the non-keratinous intermacrofibrillar matrix in the pre-keratin/

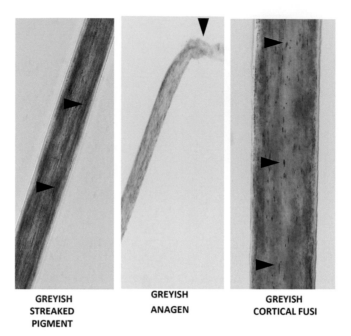

GREYISH
STREAKED
PIGMENT

GREYISH
ANAGEN

GREYISH
CORTICAL FUSI

FIGURE 4.6 (*Continued*)   (d) Greyish. Illustrated here are microscopically "greyish" coloured hairs. These light micrographs depict an actual greyish colour not to be confused with "grey hair" which is, in reality, the visual perception of darker hairs mixed with lighter or white hairs making the head of hair appear grey. Using the term greyish indicates that the hair is not completely grey but has more grey pigmentation than any other contributing pigment. Streaking (arrowhead), an anagen root (arrowhead) and cortical fusi (multiple arrowheads) can also be observed within these hairs. (*Continued*)

keratogenous region of anagen hairs. Tafaro (2000) reports on the value of microscopic post-mortem changes in event reconstruction in two murder cases.

Tip type is the final feature assessed at the LPM level of examination. Hair tips can be described in several ways. In our scheme we classify tips as showing a *natural taper, cut, rounded, frayed* or *abraded, split, crushed* or *broken* and *singed*. A single hair may demonstrate more than one of these features. For example, a scalp hair may have a natural taper and have a rounded point at the very end of the hair shaft. Examples of these features are shown in Figure 4.9.

Additional information may also be observed such as the presence of lice and disease conditions. Table 4.4 lists some of the disease conditions

FIGURE 4.6 (*Continued*) (e) Reddish. These light micrographs show a range of reddish hairs from deep red through to orangey red. Illustrated here also are two adjacent hairs from the same person indicating, to a small degree, the range of colours that are not only natural variation seen within one person's hair but often observed during an examination using LPM or TLM. The smaller arrowheads indicate the medulla in the hairs, whilst the larger arrowheads are showing smooth and coarse cortical texture. Pigment granules in red hair are generally very fine in size, shape and aggregation, as shown in Figures 4.13, 4.14 and 4.15, respectively. (*Continued*)

affecting the hair shaft. Chapter 7 "Defects of the hair shaft" from Rook and Dawber (1982) is an excellent resource providing in depth information about diseases causing defects in hair.

A schematic diagram of hair disease conditions is shown in Figure 4.10 with some examples in Figure 4.11. Many of these conditions will only be rarely seen by forensic examiners with *trichonodosis* and *trichorrhexis* perhaps being the two most commonly seen conditions.

For many organisations/laboratories this level of examination will be the final stage for hair examination. Many useful case-related questions can be answered at this level, hairs selected for DNA analysis and clearly different hairs eliminated from further examination. Hence, we see this level of examination as a triage and elimination process and not one of inclusion except for DNA analysis or more detailed microscopic examination.

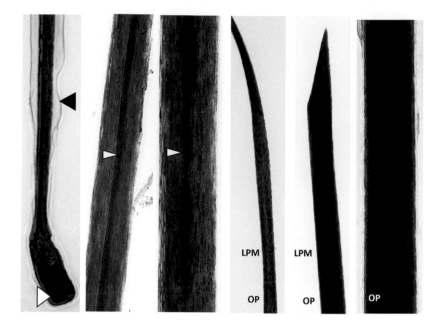

FIGURE 4.6 (*Continued*)   (f) Black/opaque. These images are light micrographs taken at both LPM and TLM and show how the coarseness of the hair (in microns) and heavy pigmentation results in either difficult to decipher features or opacity (OP) where nothing can be seen of the inherent cortical features. The combination of a thick hair and lots of pigmentation means the hair examination ends immediately upon describing the hair as opaque (OP). Some features such as the continuous medulla (small white arrowheads) are observable, whilst the sheath material (black arrowhead) and the anagen root (larger white arrowhead) can be readily seen but that is where this examination ends.

4.4 LEVEL 3 EXAMINATIONS— DETAILED EXAMINATION OF HAIRS AND COMPARISON MICROSCOPY

The examination of mounted hairs using higher magnification TLM should only be undertaken when recovered hairs have not been shown to be different and eliminated from further consideration. Depending on the case circumstances such detailed examinations may also be delayed until the results of DNA testing are available as detailed microscopic examination of hair is potentially the most time consuming.

Detailed examinations should only be conducted when there is a good reason to do so with a genuine potential for useful information to be gained from such examinations.

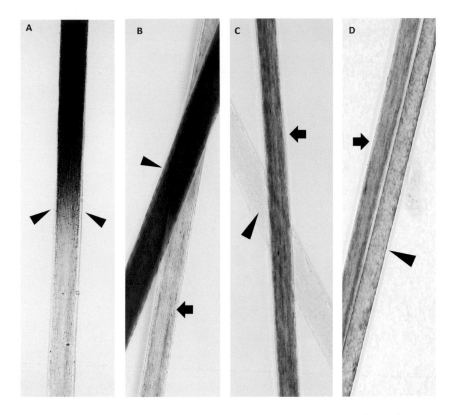

FIGURE 4.7   Dyed hair. (a) This hair originally yellow in colour shows a distinct a dye line (arrowheads) further up the hair shaft. As the approximated rate of hair growth is known it is possible to also approximate the time elapsed between an artificial colour change and the date of recovering a questioned hair. (b) These overlayed hair shafts—red (arrowhead) (hair has been dyed red) and brown (arrow) (natural hair colour) are adjacent hairs from the same scalp. As such the two hairs would provide a good basis for elimination/inclusion if they were hairs collected from a crime scene as questioned or known hairs of suspect/victim. (c) This image shows two hairs—an apparently colourless hair (arrowhead) overlying a naturally coloured brown hair (arrow). (d) Illustrated here are a naturally coloured brown hair (arrow) and a blue hair shaft (arrowhead), however, the difference between images (b) and (d) are that the blue hair is not dyed. The two hairs shown here have been tested for chemical bleaching (see Appendix 4.5). The blue hair has absorbed the methylene blue as part of a chemical reaction, whilst the unbleached hair—the brown one, shows no reaction. (c) and (d) *are four hairs from the same person.*

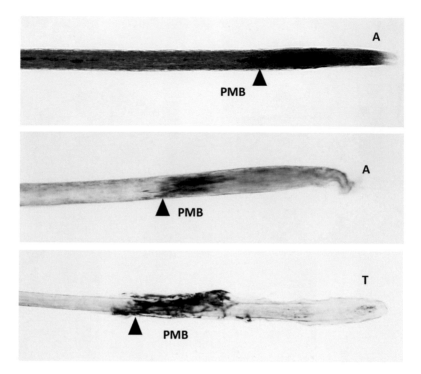

FIGURE 4.8   Post-mortem banding. Light micrographs show location of post-mortem banding (PMB, arrowhead) on both anagen (A) hairs and a telogen hair (T).

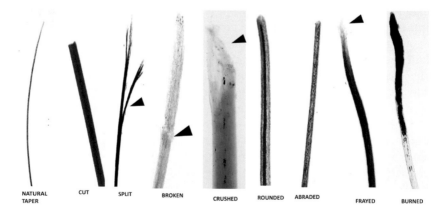

FIGURE 4.9   Hair tip ends. Different types of hair tip ends include natural taper, cut, split (arrowhead), broken (arrowhead), crushed (arrowhead), rounded, abraded, frayed (arrowhead) and burned. Other examples of tip ends can be found in Figure 2.9.

TABLE 4.4   Relationship between Diseases and Hair Abnormality

Hair shaft abnormalities
  a. Structural defects of the shaft **with** increased fragility
    i. Beaded swelling shaft
      a. Monilethrix (including pseudo-monilethrix)
      b. Trichorrhexis invaginate (bamboo hair, Netherton's syndrome)
      c. Trichorrhexis nodosa
    ii. Twisted shaft
      a. Pili torti
      b. Kinky hair (Menkes kinky hair syndrome)
    iii. Normal shaft
      a. Trichoschisis
      b. Trichothiodystrophy
      c. Trichoptilosis
      d. Structural defects **without** increased fragility
    iv. Twisted shaft
      a. Spun glass hair (uncombable hair, pili trianguli)
      b. Trichonodosis
    v. Normal shaft
      a. Pili annulate (ringed hair)
      b. Cartilage hair hypoplasia

Table 4.4 adapted from Seta, S, Sato, H and Miyake, B 1988, "Forensic hair investigation", in Maehly, A & Williams RL (eds), *Forensic Science Progress*, Volume 2, New York: Springer-Verlag Berlin.

    The examiner must have access to a good quality TLM with a range of objectives up to and including at least a ×40 dry objective. Recovered/questioned hairs and known hairs can be examined using a stand-alone microscope. All comparisons require a high-quality comparison microscope, such as that shown in Figure 4.12.

    There is some debate about the sequence or order that should be taken in examining hairs with the aim of avoiding **context bias**. Best practice would indicate that recovered hairs should be first examined to avoid the examiner "finding" features that they will have already seen in known samples. As detailed examination very much relies on visual observation it is always possible that the examiner is drawn to an obvious feature. In some case circumstances there should be no problem with taking this approach. However, in some circumstances it is necessary to examine a known sample first. For example, where the body of a deceased person is found, a known sample obtained at post-mortem, and where subsequent samples are collected to be compared to that known sample from one or more suspects and sometimes over a long period of an ongoing investigation.

INCREASING FRAGILITY

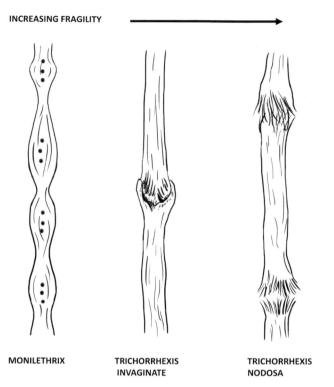

MONILETHRIX   TRICHORRHEXIS   TRICHORRHEXIS
INVAGINATE   NODOSA

FIGURE 4.10   (a) Schematic diagram of hair disease conditions. *Monoilethrix*: Trichoscopy (a method of hair and scalp evaluation that is used for diagnosing hair and scalp diseases) classifies monoilethrix as a "shaft narrowing" abnormality (Rudnicka *et al*, 2012). It is a rare hereditary condition generally considered to be an autosomal-dominant disorder with variable penetrance (Baltazard *et al*, 2017), where the hair shaft has visible, regularly distributed nodes and internodes. These nodes correspond to normal hair shaft thickness, whereas the internodes are narrowings. Hair shafts are bent in different directions and tend to fracture at constriction points. Often referred to as "beaded or necklace" hair. *Trichorrhexis invaginate*: Often referred to as "bamboo hair" trichorrhexis invaginate is classified as having a node-like appearance (Rudnicka *et al*, 2012). Bamboo hairs are characteristic of Netherton's syndrome (Netherton, 1958) that is a very rare genodermatosis characterised by ichthyosis (genetic skin disorders characterised by dry, thickened, scaly skin) and hair shaft abnormalities. In trichorrhexis invaginata, the hair shaft telescopes into itself. The proximal part of the abnormality is concave, whereas the distal end is convex (bulging). This produces a nodular swelling along the hair shaft with a "ball in a cup" appearance. The presence of many of these nodes along the hair shaft gives it the appearance of bamboo. *Trichorrhexis nodosa*: In trichorrhexis nodosa, the hair develops a restricted area in which the shaft splits longitudinally into numerous small fibres. The outer fibres bulge outward, causing a segmental increase in hair diameter and are classified by Rudnicka *et al* (2012) as having a node-like appearance. At high magnifications and in dry trichoscopy, as seen in Figure 4.11, the split hair shaft appears as two brown brushes aligned in opposition *(arrowheads)*. (*Continued*)

SHAFT ABNORMALITIES – STRUCTURAL DEFECTS - TWISTED SHAFT

INCREASING FRAGILITY ➤

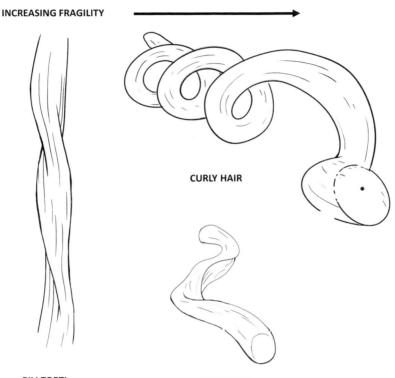

CURLY HAIR

PILI TORTI   KINKY HAIR

FIGURE 4.10 (*Continued*)   (b) Schematic diagram of hair disease conditions. *Pili torti:* Classified by Rudnicka *et al* (2012) as part of the group of hair shafts with curls and twists, hair shafts affected by pili torti, are characterised by hair that is flattened and twisted on its own axis at irregular intervals, usually at a 180° angle. Pili torti has numerous causes, both inherited and acquired. *Curly hair:* The curly hair fibre appears to be a distinct type of fibre, with its own physical, mechanical and biological make-up (Cloete *et al*, 2019), where the anatomy of the curly hair forming within the follicle produces a curved fibre that has flattened areas (in cross-section) along the hair shaft that allows it to curl. *Kinky hair:* Kinky hair is a characteristic of Menkes disease (Menkes *et al*, 1962) where the appearance of the hair is poorly pigmented, woolly, sparse and friable with easy pluckability (Datta *et al*, 2008). This disease is an X-linked lethal multisystem disorder caused by disturbances of copper distribution in different tissues (Datta *et al*, 2008). (*Continued*)

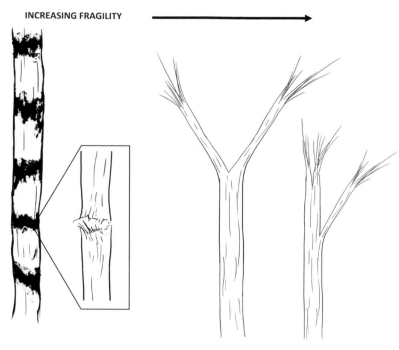

SHAFT ABNORMALITIES – STRUCTURAL DEFECTS - NORMAL SHAFT

INCREASING FRAGILITY

TRICHOSCHISIS   **TRICHOPTILOSIS**

FIGURE 4.10 (*Continued*)   (c) Schematic diagram of hair disease conditions. *Trichoschisis*: Classified as part of the fractured hair shaft abnormalities (Rudnicka *et al*, 2012) trichoschisis is characterised by a clean
transverse fracture across the hair shaft that results from absence of the
hair shaft cuticle in the affected area. Trichoschisis is a common finding in trichothiodystrophy the latter being a rare autosomal recessive
inherited disorder characterised by sulphur deficient brittle hair, intel-
lectual impairment and multisystem abnormalities (Rudnicka *et al*,
2012). *Trichoptilosis*: Also classed as a fractured hair shaft abnormality, trichoptilosis is commonly known as "split ends" and refers to the
longitudinal splitting of the distal end of the hair shaft. Split ends are
not a pathognomic condition generally resulting from grooming over
time, hair styling and cosmetic procedures, and environmental factors
(Rudnicka *et al*, 2012). (*Continued*)

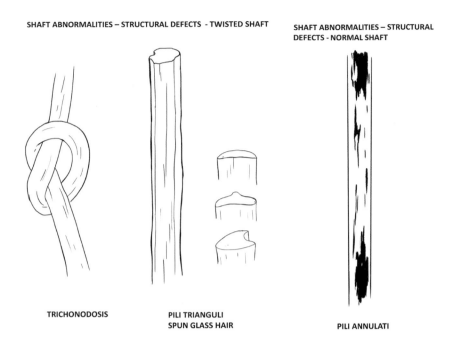

TRICHONODOSIS

PILI TRIANGULI
SPUN GLASS HAIR

PILI ANNULATI

FIGURE 4.10 (*Continued*) (d) Schematic diagram of hair disease conditions. *Trichonodosis*: Classed as part of the hair shaft abnormalities with node-like appearance, this condition is one of the more commonly occurring ones also illustrated in Figure 4.11. Commonly known as "knotted hair" it is an acquired, transient condition in which a single or double knot occurs in the hair shaft, either spontaneously or in response to scratching. It is observed in both long and short hair. Trichonodosis is usually an incidental finding of little clinical significance (Rudnicka *et al*, 2012). *Pili Trianguli*: Also known as uncombable hair syndrome or "spun glass hair" is a rare autosomal dominant genetic disorder. It usually presents in early childhood with straw-like, silvery blond unruly hair that completely resists all efforts to control it with a brush or comb. Hair density and quantity are within normal ranges, and fragility is rare. This condition is readily diagnosed because of the typical kidney-like or triangular appearance of hair in cross sections. The longitudinal grooving or the less common triangular hair shape renders the hair rigid and could explain the physiopathology of this rare condition (Piccolo *et al*, 2018). *Pili annulati*: These hairs are part of the group hairs that have "banded" shafts defined by characteristic alternating light and dark banding due to air-filled spaces between the macrofibrillar units of the hair cortex. Not essentially a fragile hair but affected hairs are more prone to weathering particularly in combination with androgenetic alopecia (Hofbauer *et al*, 2001).

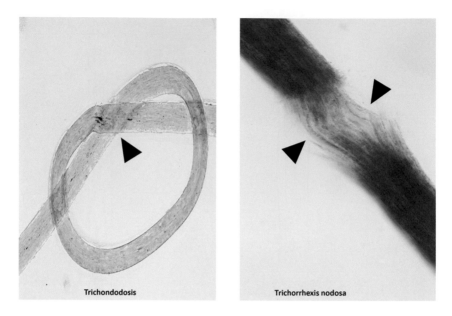

Trichondodosis   **Trichorrhexis nodosa**

FIGURE 4.11   Hair disease conditions. Light micrographs show two conditions occasionally observed during hair examination. Whilst these two conditions are by no means common, they are certainly more often seen than many of the other conditions as illustrated and described in Figure 4.10. Trichonodosis, often called knotted hair, are single strands of hair that knot within themselves (arrowhead) to create bigger knots. Curly and kinky hair are more prone because these hairs have a flat and curvy hair profile which encourages the knots to form and tangle. *Trichorrhexis nodosa* is seen as longitudinal fractures occurring along the hair shaft (arrowheads) and are often caused by environmental factors including perming, harsh grooming, blow drying and exposure to chemicals. The condition is self-correcting when environmental and grooming practices improve.

In our view, the strict use of checklists to record the examiners observations limits the risk of unconscious bias. Furthermore, our approach to hair examination is that we look for differences with a focus on elimination. It is only at the final comparison microscopy stage that we introduce the concept of "no meaningful differences" in then deciding the next step of reaching a potentially inclusionary finding.

4.4.1 High-Power Transmitted Light Microscopic Examination

It is important to stress that examination of microscopic features must always include examination of hair shafts along their entire length from

FIGURE 4.12   Modern transmitted light comparison microscope for assessing samples side by side.

root to tip end paying attention to the pattern of features observed as well as recording the detail of individual features. In general, the features seen at this level of examination are associated with the cuticle, the cortex and the medulla. Information about the root and tip ends will be obtained which may add to or complement the descriptions obtained with the use of LPM.

Robertson (1999b) discusses the development of the feature classification used by the authors (see checklist; Appendix 4.2b). This checklist was developed from the study of Robertson and Aitken (1986) and those features agreed by the 1980s Committee on Forensic Hair Comparison (CFHC) (Anon, 1985a). As we have stressed on numerous occasions the use of a checklist ensures a systematic, thorough and comprehensive examination process. In developing the checklist, the aim was to balance several competing factors. These include the following:

- Too many subcategories for each feature may lead to increased subjectivity.
- Too few subcategories may miss useful information.
- Too many subcategories may lead to the examiner being overloaded with potential information—they may then not record information accurately because it is too onerous or may miss a pattern because they are too focused on detail.

We recognise that a potential limitation of microscopic examination is the subjective nature of capturing information based on observation as by definition "observation" is examiner dependent. The challenge is to ensure a high level of consistency when the same observer examines a hair on more than one occasion and, as far as possible, to obtain consistency in recording between observers examining the same hair.

It is not currently possible to capture microscopic features in a truly numerical sense. This is not to say that there are no microscopic features that would not be capable of being measured but a number of historical studies have failed to demonstrate that those features that can be measured have any ability to discriminate or differentiate between hairs from different individuals (Robertson, 1999a). Many of the measurable features have looked at various ways of measuring the cuticle scales. However, these were almost certainly doomed to be of little value as the inherent scale pattern for human hairs shows no meaningful variation between individuals.

To develop some meaningful metrics for hairs we have previously assessed the use image analysis to capture microscopic features with the aim of addressing the issue of the subjective nature of observational classification of features (Brooks *et al*, 2010). Some of the challenges that need to be overcome, if such an approach is to ultimately prove to be of value, include the following:

- At magnifications high enough to see microscopic features (potentially up to a magnification of ×1,000) the depth of field will be very small—in order to obtain a fuller appreciation of some features such as pigmentation it is necessary to look at a number of optical sections using a technique such as auto montage software.
- Algorithms would need to be developed to look at pigment patterns in a holistic way and derive some form of classification.
- Data capture and analysis would have to be repeated many times for each hair.

Robertson (1999b) hoped that it would only be a matter of time before these challenges would be overcome and some form of expert system based on machine learning would be developed to assist, not replace, the hair examiner. Some 20 years later we are still waiting on a useful and practical application along these lines. To a large extent the ability to conduct DNA analysis has reduced the emphasis on microscopic examination of hairs, but for those organisations and individuals who still conduct high magnification examinations of hairs an expert system would still be a welcome addition to their armoury.

Accepting that we must currently rely on the observer deciding on the microscopic features they see the microscopic checklist (Appendix 4.2b) attempts to reach the right balance of the competing factors outlined earlier in this section by limiting the number of subcategories in each of the various features.

The first feature recorded is hair **shaft diameter**. This should be recorded at various points along the hair shaft, the number of points largely depending on the length of the hair and observed variation of hair shaft diameter.

The next group of features relate to **hair pigmentation**. In Chapter 2, we discussed the formation of pigmentation in the hair shaft and we have also previously discussed the visual assessment of hair colour which has a direct relationship to the underlying pigmentation. Pigmentation is subdivided into six categories these being *pigment density, pigment distribution (across the hair shaft), pigment granule size, pigment granule shape, pigment aggregate size and pigment aggregate shape*. The latter two categories recognise that in some hairs pigment granules are seen in recognisable groups or aggregates and then attempting to classify these groups based on the size and shape of the clumps or aggregates.

The subcategories for pigment features are reasonably self-explanatory. Examples of each pigment feature and subcategories are shown in Figures 4.13–4.18.

It should be stressed that size of pigment granules and aggregates is not based on absolute measurements and is a comparative measure assessed relevant to the hair samples in each case. The size of individual pigment granules will be at the very limits of resolution even when examined with a ×40 objective with a high numerical aperture.

When assessing streaked aggregation some caution needs to be exercised not to confuse this with underlying coarse cortical texture which may not be readily visible in a heavily pigmented hair.

Other features that may be observed in the cortex include *ovoid bodies (OB), cortical fusi and cortical texture*.

OB are well defined, highly dense clumps of pigmentation. Their origin may be as undispersed melanosomes. The presence of OB is not rare, but neither are they always present. Where they are present their abundance can vary widely from very few and disperse to numerous and widely distributed along the hair shaft (see Figure 4.19). There are no published studies that give an insight as to how discriminating is the presence and abundance of OB although personal experience would indicate that this can have discrimination value in cases.

Cortical fusi are spaces between cortical cells which become elongate or fusiform as cortical cells elongate during cell differentiation and maturation in the follicle. They are readily visible under TLM examination and are most often found towards the root end of the hair shaft

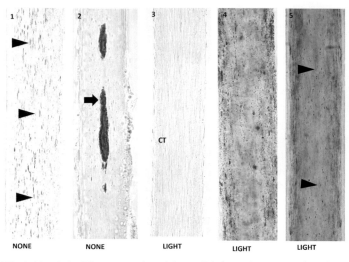

FIGURE 4.13 (a) Pigment densities—Light pigment density. Light micrographs of scalp hairs showing *no* pigment density and *light* pigment density. 1 & 2. No pigmentation (aka white hair). The arrowheads indicate cortical fusi, not pigment granules and the arrow indicates an intermittent medulla. 3. Lightly pigmented pale yellow (aka blond) hair that has a coarsely textured shaft (CT). 4 & 5. Two lightly pigmented brown hairs, with cortical fusi (arrowhead) in the shaft of hair 5.

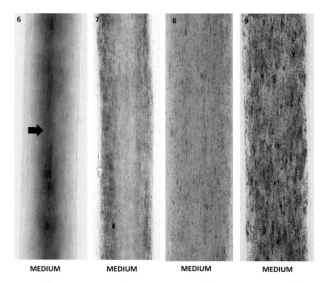

FIGURE 4.13 (b) Pigment densities—Medium pigment density. Light micrographs showing *medium* pigment density. 6. Red hair with medium density pigmentation and almost continuous medulla (arrow). 7. A yellow brown hair shaft with medium density pigmentation trending towards the cuticle. 8 & 9. Two other examples of medium density pigmentation in different brown-coloured hairs. (*Continued*)

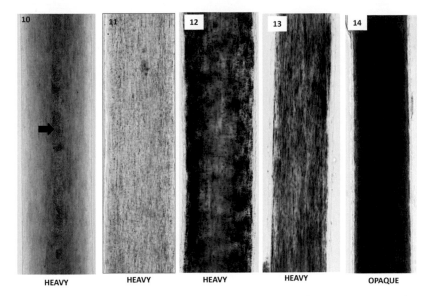

| HEAVY | HEAVY | HEAVY | HEAVY | OPAQUE |

FIGURE 4.13 (*Continued*) (c) Pigment densities—heavy pigment density. Light micrographs showing *heavy and opaque* pigment density. 10. Red Hair with heavy density pigmentation and almost continuous medulla (arrow). The pigment granules in this red hair are extremely small and fine and the depth of the red colour is indicated by a heavy density of pigmentation. 11, 12 & 13. Shades of brown hairs with variations of heavy pigmentation. 14. Opaque—where the pigmentation is so dense that light cannot pass through.

(see Figure 4.20). Care needs to be taken not to miss their presence in heavily pigmented hair and they should not be confused with hair pigmentation. As well as their presence their distribution along the hair shaft can provide additional discrimination between hairs from different individuals.

The only feature that relates directly to the cells forming the cortex is that of **cortical texture** (Figure 4.21). Under TLM, it is rarely possible to see the outlines of individual cortical cells as they are closely packed into a rigid and homogenous hyaline mass (Seta *et al*, 1988). In hairs from a normal healthy individual with undamaged hair, it is uncommon to see coarse cortical texture. It is likely that where the outline of cortical cells is visible enough to be assessed as coarse this is the result of some physical or chemical disruption. Not surprisingly cortical texture is often seen towards the tip end of hairs that have been the subject of repeated cosmetic treatment. Cortical texture may be missed in more heavily pigmented hairs and is most often seen in hairs with less dense

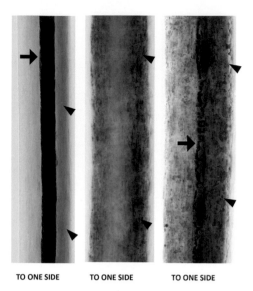

TO ONE SIDE   **TO ONE SIDE**   **TO ONE SIDE**

FIGURE 4.14 (a) Pigment distribution to one side of hair shaft. In each of these light micrographs there is an apparent concentration of pigmentation towards the right-hand side of the hair shaft (arrowheads), and whilst still pigmented on the left-hand side there is distinctly less pigment (colour) deposited. There are continuous medullas (arrow) in two hairs (the red hair and the darker brown hair).

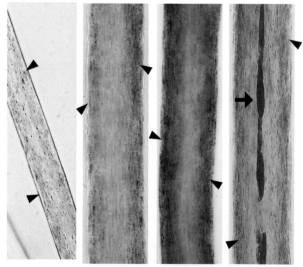

TOWARDS CUTICLE   **TOWARDS CUTICLE**   **TOWARDS CUTICLE**   **TOWARDS CUTICLE**

FIGURE 4.14 (b) Pigment distribution towards the hair cuticle. These images illustrate pigmentation that tends to be toward the cuticle (arrowheads) leaving a distinctly lighter zone in the centre of the hair shaft. One hair has a non-continuous medulla (arrow). (*Continued*)

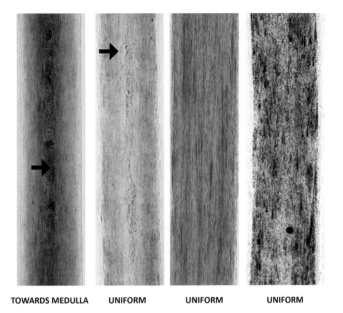

TOWARDS MEDULLA **UNIFORM** **UNIFORM** **UNIFORM**

FIGURE 4.14 (*Continued*) (c) Pigment distribution towards the medulla and uniform distribution. Red hair generally has pigment distribution *towards the medulla* (arrow) as illustrated by the deeper colouration in the central area of this hair image. *Uniform* pigment distribution shown in the other three images (the lighter brown hair has a translucent medulla—arrow), that is readily observed in hairs at both LPM and TLM.

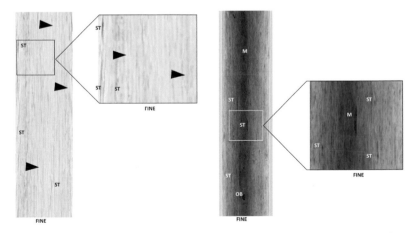

FIGURE 4.15 (a) Pigment granule *sizes*—fine. Two sets of examples showing fine sized pigment granules. In microscopic terms "fine" here means that individual granules cannot be resolved at TLM. In some areas of the hair shafts where the fine pigment granules accumulate to form an aggregate or streak (ST) and in other areas undispersed melanin forms ovoid bodies (OB). The red hair has a remnant medulla (M). The yellow hair (aka blonde) is characterised by overall cortical texture (arrowheads). (*Continued*)

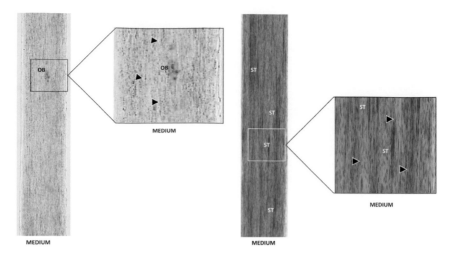

FIGURE 4.15 (*Continued*)   (b) Pigment granule *sizes*—medium. Two sets of examples showing medium-sized pigment granules. The difference between the fine and the medium granule sizes is again related to the ability to resolve an individual granule (arrowheads) against the general background of cortical colour. Some small ovoid bodies (OB) are apparent in one hair and many streaks (ST) of pigment in the other hair shaft.

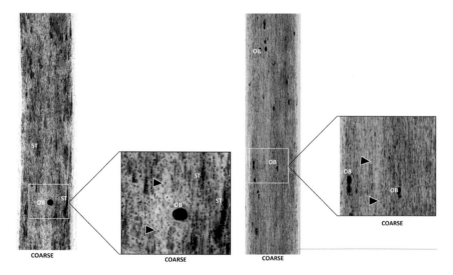

FIGURE 4.15   (c) Pigment granule *sizes*—coarse. Two sets of examples showing coarsely sized pigment granules. Coarse granulation is readily seen at TLM as is clear from the unexpanded images. Ovoid bodies (OB) are sometimes individual and large or multiple and smaller. Streaking (ST) seen here is concentrated enough to form aggregates or clumps of pigment—all of which add to the patterning of the hair with its own distinctive characteristics.

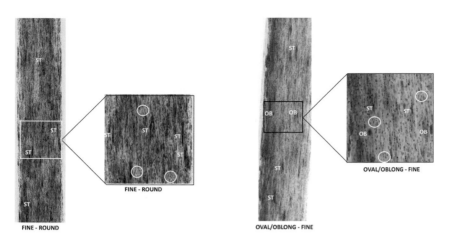

FIGURE 4.16 (a) Pigment granule *shapes*—fine—round/oval/oblong. These two sets of light micrographs illustrate both the size and the shape of the pigment granules. The fineness or size of the granules are larger (visible) than those in Figure 4.15 and have distinct round or oval shapes (within the white circles). Where the pigment granules aggregate, they form fine streaks (ST). As distinct from the streaks are the appearance of ovoid bodies (OB) in one hair.

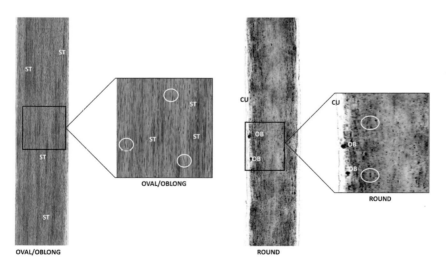

FIGURE 4.16 (b) Pigment granule *shapes*—oval/oblong/round. The pigment granule shapes shown here are slightly larger and therefore more easily distinguished in the images (within the white circles). Again streaking (ST) is observed as are the presence of ovoid bodies (OB).

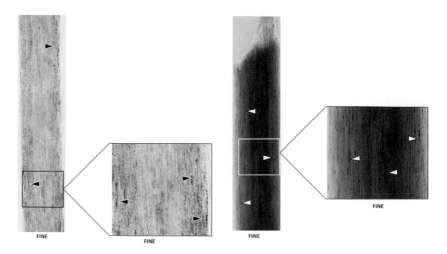

FIGURE 4.17 (a) Pigment aggregate *sizes*—fine. The shape, size and "density" of aggregate shapes present within the cortex are often distinctive characteristics that allow the separation of one person's hair from another's. Shown here are two examples of aggregations that occur as fine streaks (arrowheads). It is useful to compare the three parts that make up Figure 4.17 to see how the aggregate sizes increase in density, depth and physical entities.

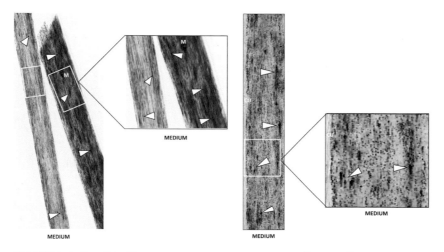

FIGURE 4.17 (b) Pigment aggregate *sizes*—medium. Illustrated here these light micrographs show medium sized aggregates (arrowheads) an intermittent medulla (M) and the cuticle (CU). Also obvious are the individual round pigment granules in the right-hand hair. (*Continued*)

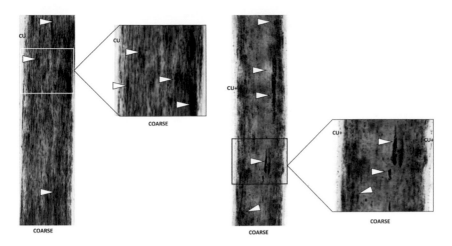

FIGURE 4.17 (*Continued*)   (c) Pigment aggregate *sizes*—coarse. large, coarse aggregations (arrowheads) of pigment seen here occurs in hairs with coarser pigment granulation (right hair) and fine pigment granulation (left hair). Cuticle can be observed in both hairs (CU) and (CU+) where the latter shows pigment granules within the cuticle layers.

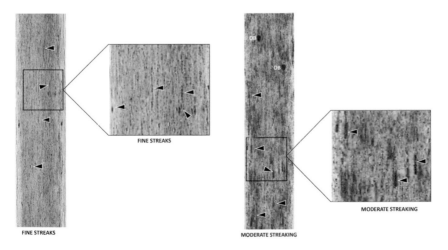

FIGURE 4.18   (a) Pigment aggregate *shapes*—fine streaks. Arrowheads indicate both fine and moderate shaped streaks of pigment within the two hairs seen here. Ovoid bodies (OB) are also observed. (*Continued*)

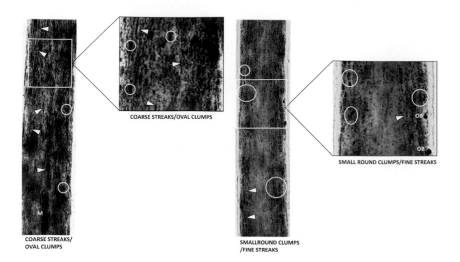

COARSE STREAKS/OVAL CLUMPS

SMALL ROUND CLUMPS/FINE STREAKS

COARSE STREAKS/
OVAL CLUMPS

SMALLROUND CLUMPS
/FINE STREAKS

FIGURE 4.18 (*Continued*)   (b) Pigment aggregate *shapes*—coarse streaks, oval clumps and small round clumps. The aggregate shapes illustrated here include coarser streaks (arrowheads) than those seen in Part (a) of Figure 4.18; oval clumps of aggregated pigment (within the white circles); and small round aggregate clumps (within the white circles) associated with fine streaks (arrowheads). Ovoid bodies (OB) are also present. (*Continued*)

pigmentation. Care should be taken to not confuse cortical texture with streaked pigmentation as a visual impression of streaking can result from underlying cortical texture.

Examples of cortical features are shown in Figures 4.19–4.21.

As discussed in Chapter 2 human hairs may or may not have a visible central core or **medulla** which occurs when cells collapse as cell membranes break down and dead cells dehydrate. Where present and visible the medulla in human hair occupies less than one third of the width of the hair shaft. The ratio of the width of the medulla to the width of the hair shaft is called the *medullary index* (MI) and whilst this index is useful to differentiate animal and human hairs, it has no value in discriminating between the source of human hairs. MI has also been shown to have no correlation with race or sex of an individual (Seta *et al*, 1988).

By comparison to the appearance of medulla structure in animal hairs, the medulla in human hairs is unremarkable being amorphous (or unstructured) in appearance. The appearance of the medulla as seen under TLM is often described as *translucent* or *opaque* and this is the terminology used in our checklist. It is important that the observer understands the underlying cause of this difference in appearance. A translucent medulla results when air trapped in the intercellular and

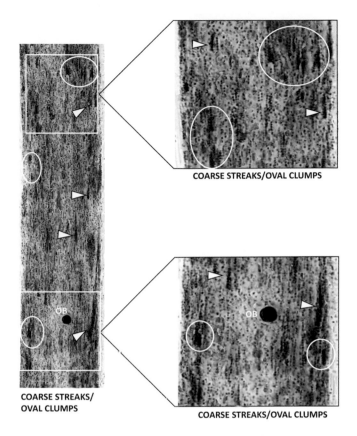

FIGURE 4.18 (*Continued*) (c) Pigment aggregate *shapes*—coarse streaks and large oval clumps. Observed in this single hair are coarse streaks (arrowheads); numerous oval clumps (within the white circles) that are in themselves coarse and a large, distinctive ovoid body (OB).

intracellular gaps in the medulla is displaced by mountant used for microscopy. Where the air is not displaced the medulla will appear dark and is classified as opaque. It is our view that the appearance of the medulla is not an especially useful feature as it is most likely an artefact rather than an inheritable feature of hairs. This feature is included only because it is a visual reality for the observer and needs to be explained.

The second feature of the medulla that needs to be assessed is its occurrence along the length of the hair shaft. The observer needs to ensure that they do not miss the presence of sometimes less easily seen translucent sections of medulla. Examination of hairs using polarised light can help visualise the full extent of the medulla. Various terms have been used in studies to classify the extent of the medulla with terms such

FIGURE 4.19   (a) Ovoid bodies (OB). Hairs (A–E) illustrate some of the frequencies and sizes of OB observed in scalp hairs. A & B. Small and infrequent. C. small and in chains (within the white oval). D & E. Slightly larger OB but still infrequent. The frequency and size of the OB illustrated here would be noted but not seen as a feature apart from the chain seen in hair C. If a number of chains occurred along the hair shaft this could be noted as a distinguishing characteristic. (*Continued*)

as absent, broken, continuous, fragmentary and interrupted having been used. In our checklist we classify the medulla in a very simple way as, *none or not visible, medulla occupying less of the hair shaft than the area not showing a medulla (medulla < space), medulla occupying more of the hair shaft than the area not showing a medulla (medulla > space), or continuous* where the medulla is present along all or almost all of the length of the hair shaft. Even in hairs with a continuous medulla it may not be present at the very tip end or close to the root end.

In general, the medulla is more prominent, has a higher MI and is likely to be closer to continuous in non-scalp hairs than in scalp hairs.

A rare occurrence is the presence of a double medulla (Montagna and Van Scott, 1958).

Examples of medulla appearance in human hairs are given in Figures 4.22 and 4.23.

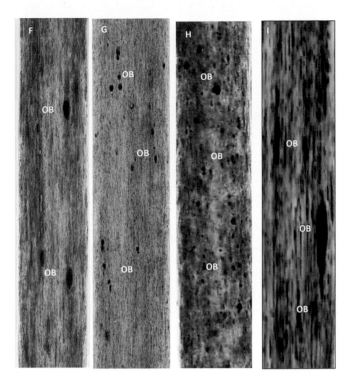

FIGURE 4.19 (*Continued*) (b) Ovoid bodies (OB). Hairs (F–I) feature OB that would be seen as a very distinct characteristic of each of these hairs. F. Large-sized bodies. G. Small size but frequent. H. Combination of frequency, shape and regular distribution. I. Very large size of the ovoid body.

The last features that are assessed in our scheme are those relating to the appearance of the **cuticle**. These include the colour and thickness of the cuticle and the appearance of the scale layer comprising the cuticle.

The cuticle will normally be either colourless or showing a slight yellow hue. Normally the cuticle will not have pigment granules present so when seen these should be noted as present. The cuticle can also show some variation in thickness and this should be recorded.

As previously described the outer layer of the hair shaft, or cuticle, is composed of flattened-scale-like cells which overlap like roof tiles. The cells overlap longitudinally and laterally to envelop the cortex and protect it from environmental exposure. The scales slope outwards and their free edges point towards the tip end of the hair shaft. The cuticle of a mature human hair is on average six cells thick, each overlapping. It has been known since studies in the 1920s (Hausman, 1925) that the pattern

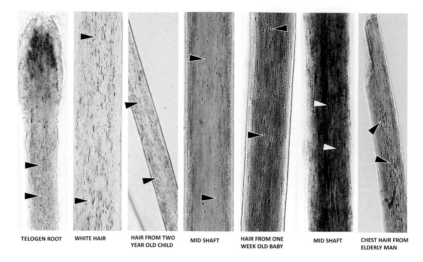

TELOGEN ROOT | WHITE HAIR | HAIR FROM TWO YEAR OLD CHILD | MID SHAFT | HAIR FROM ONE WEEK OLD BABY | MID SHAFT | CHEST HAIR FROM ELDERLY MAN

FIGURE 4.20   Cortical fusi. Cortical fusi frequently occur just above the root of telogen hairs where pigmentation and cell production has ceased. They are also frequently seen in colourless (white) hairs, and can occur at both root end and along the hair shaft. As it is more uncommon for cortical fusi to be found in numbers along the full length of the hair shaft it is a characterising feature and should be noted.

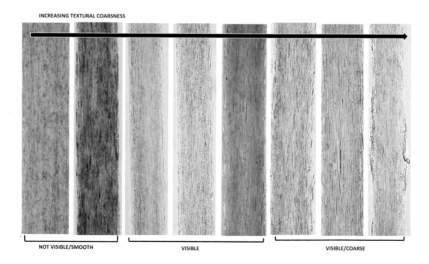

INCREASING TEXTURAL COARSNESS

NOT VISIBLE/SMOOTH | VISIBLE | VISIBLE/COARSE

FIGURE 4.21   Cortical texture. The presence of cortical texture can sometimes be misinterpreted as pigment streaking and can vary in its appearance from smooth or not visible to being visible to coarse and obvious. The latter is often observed in very damaged hair caused by environmental, chemical treatments such as artificial colouring, perming, etc., and/or health conditions.

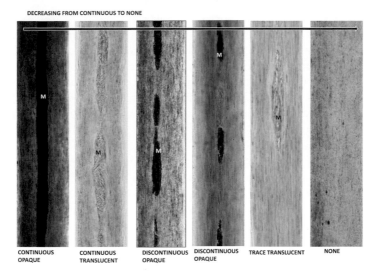

FIGURE 4.22 Medulla distributions. The presence of a medulla in human hair can vary from being continuous to none at all and can be either opaque (air filled) or translucent (liquid filled from the mountant) when observed by microscopy. The medulla may or may not be seen at LPM but will be very obvious (for its presence or absence) at TLM.

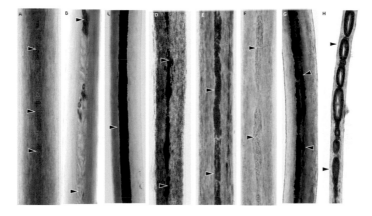

FIGURE 4.23 Medulla types. (a) In this red hair the medulla is present (arrowheads), discontinuous and not very obvious. The pigment in red hair is usually denser towards the medulla. (b) A continuous medulla mainly translucent (arrowhead) with some opaque regions (arrowhead). (c) Continuous opaque medulla (arrowhead). (d) Discontinuous medulla (arrowheads) with some apparent fragmentation. (e) Continuous medulla (arrowheads) with some apparent fragmentation. (f) Translucent, discontinuous (arrowheads) medulla. (g) An example of a double medulla (arrowheads) generally found in beard hairs but sometimes occurs in scalp hairs. This is a beard hair. (h) A bubbled medulla (arrowheads) indicating this hair has been exposed to high temperatures.

of scales in human hair do not show any significant natural variation between individuals and, hence, has no value in discriminating between individuals. However, the cuticle and scales are subject to wear and tear through weathering, environmental exposure and other external factors such as cosmetic treatment including brushing, combing and bleaching and dying treatments (Robbins, 1988). Often the impact of these "assaults" is most evident towards the tip end of hairs, especially in longer hairs where the hair shaft has a longer exposure time and potential impact for these factors to play out. Where damage is extreme the cuticle can be lost completely exposing the cortical cells to the environment. This can result in frayed ends where the elongated cells comprising the cuticle can be seen at the TLM level of examination. The hair shaft may also split to give a typical appearance called *split ends* that are visible at LPM.

At the TLM level of examination, the outer margin of the cuticle is classified in our checklist as being *smooth, serrated, ragged, cracked* or *looped*. Although these features are not inherited, they are characteristics often related to lifestyle and health condition of the individual.

No mention has been made of the use of scanning electron microscopy (SEM) for the examination of human hairs and to visualise scale surface features. There is no doubt that SEM produces visually attractive images of surface features such as scale patterns. However, as previously discussed the scale pattern in human hairs shows no meaningful natural or inheritable variation between individuals and acquired characteristics are readily assessed as described above with the use of LPM. For animal hairs, scale pattern features are of great value in identifying animal hairs but these are relatively easily and cheaply seen with scale castes with the additional advantage that a caste can be made of the complete hair allowing the study of the scale pattern along the length of the hairs' shaft. Hence, SEM has no practical role to play in the normal examination of human hairs.

Images of cuticle appearance are shown in Figure 4.24.

4.4.2 Comparison Microscopy

The role of comparison microscopy will be considered in more detail in Chapter 5, where we look at examination protocols, conclusions and reporting the outcomes of hair examinations. We have repeatedly stressed the need to fully record all examinations using a combination of checklists, drawings, images and written descriptions and summaries as appropriate. If the purpose of an examination is limited to selecting hairs for DNA testing or even with LPM examinations, there will be no need to move to the comparison microscopy stage. Even using TLM of recovered or questioned hairs there may be no need for comparison microscopy where they have been shown to differ from relevant known

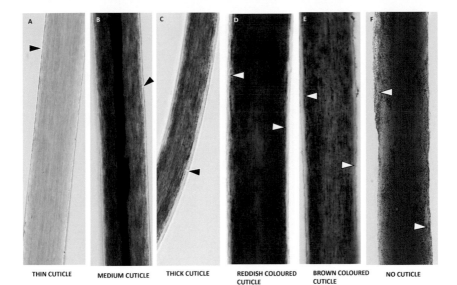

| THIN CUTICLE | MEDIUM CUTICLE | THICK CUTICLE | REDDISH COLOURED CUTICLE | BROWN COLOURED CUTICLE | NO CUTICLE |

FIGURE 4.24 (a) Cuticle types. The cuticle forms an outer protective layer of the hair that protects the softer inner core of the hair from wear and tear. Cuticle is generally clear as observed by TLM except in hairs that have been artificially coloured—for example, hairs D & E where the cuticle layer is reddish or brown, respectively. The cuticle layer is measured for thickness/width and colour are noted during examination at TLM (see Appendix 4.2b) as it forms part of the overall character detail of the particular hair. Cuticle thicknesses relates to the number of cuticle cell layers found within the hair cuticle layer. Generally, there are 5–10 scale layers in human hair where each cuticle cell can be 0.3–0.5 μm thick (Guohua *et al*, 2005). Illustrated here are A—thin cuticle; B—medium thickness cuticle; C—thick cuticle; D & E—coloured cuticle and E—no cuticle. The cuticle on this hair, E, has a dermatophyte infection called ectothrix which is a fungus that lives on keratin, eventually consuming the cuticle outer layer (Adya *et al*, 2011). (*Continued*)

samples. However, where the differences are subtle or small, it may be necessary to confirm differences between a recovered hair and a known hair with a side-by-side comparison.

In the event that the examiner is unable to eliminate a recovered hair at the TLM level of examination then it is an **absolute** requirement that the recovered hair has been directly compared using a comparison microscope. Appendix 4.2c shows a suitable form to record such comparisons. The examiner should always identify which known hair(s) have been used to compare against the recovered hair. In our view it is not

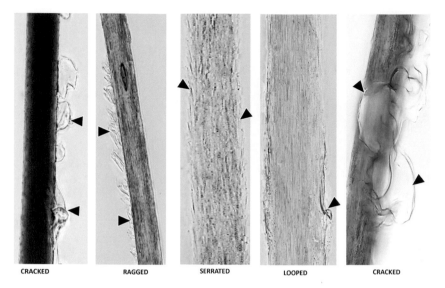

| CRACKED | RAGGED | SERRATED | LOOPED | CRACKED |

FIGURE 4.24 (*Continued*) (b) Cuticle types. In Figure 4.24(a) hair images A–E displayed a "normal" cuticle layer—that is clear or coloured, and appearing as a discreet character. Cuticle itself is damaged by processes such as grooming, environmental, washing, abrasion and chemical treatments during artificial colouring, perming or heating during straightening. Cuticle is dead tissue and unable to repair itself so over time the cuticle layers break down as illustrated above. The cuticle may crack, become ragged, serrated or looped. Such conditions should be noted as they form an "acquired" feature and part of that hairs' character list.

sufficient to simply say the recovered hair could not be eliminated against the whole of the known sample. Both internal review, and external review where conducted, should be able to repeat the same comparison as used by the primary examiner to reach a non-exclusionary finding.

Hence, the comparison process can involve multiple comparisons of each recovered hair against several known hairs. In each comparison process, the two hairs being compared should be examined along the length of respective hair shafts. In order to not exclude there must be *no meaningful differences* in the overall pattern of features and in the detailed features at several points of comparison along the hair shaft. The specific number of comparison points will depend on the length of the hairs being compared and how variable are the microscopic features. As a guide we would recommend a minimum of three specific areas of direct comparison.

Some organisations may include a requirement to capture an image of each comparison. In general, we do not favour such an approach as this can promote the presentation of the examiners findings as an image inviting non-experts to reach their own conclusion based on one image which may or may not reflect the more complex holistic information on which an expert examiner will reach a conclusion. We are not against capturing images to show specific features which may be used to explain the process when giving evidence but NOT the actual comparison process. Given the often-lengthy time between laboratory examination and the court process, capturing relevant images may also assist the examiner to recall the detail of their examinations.

As can be seen from the above brief discussion of the comparison process this can be quite onerous even for a single recovered hair. Given that many cases will involve many recovered hairs and possibly many known hairs samples, the comparison stage should not be undertaken lightly. If the examiner rejects large numbers of hairs at the comparison stage, it would tend to suggest that too many hairs have not been excluded at the TLM stage. From our experience we know that there will be cases where the differences between two known samples are so subtle that exclusions will only be reached at the comparison stage, but this should be a rare event. In most cases if a hair has not been excluded by TLM, comparison microscopy will confirm that an exclusion based on microscopic appearance is not possible. This then takes the examiner to the next stage of evaluating the weight of a non-exclusion. We will turn our attention to this in the next chapter.

Figure 4.25 shows some examples of hair comparisons.

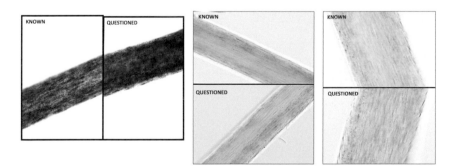

FIGURE 4.25 Comparison of known and questioned hairs in the comparison microscope. Illustrated here are three light micrographs showing the comparison between the known and the questioned hair. Important to note is the light balance in both microscopes is as close as possible otherwise the hair colour will appear different. In the dark brown hair, the cortical features—pigment streaking and aggregates and the colour suggest an inclusion. For the other two examples it is the cortical texture, the condition of the cuticle and the location of the pigmentation that suggests inclusions.

APPENDIX 4.1A: ANIMAL HAIR
EXAMINATION PROFORMA

NATIONAL CENTRE FOR FORENSIC STUDIES NCFS UNIVERSITY OF CANBERRA		Page		of		
ANIMAL HAIR EXAMINATION RECORD—Sheet 1 Case Reference:						
Hair Number		1	2	3	4	5
Macroscopic Feature		R➜T	R➜T	R➜T	R➜T	R➜T
Length	CM					
Shaft Profile	Shield					
	Straight					
	Symm. thick					
	Wavy					
Colour						

General description and comments:

Examined by:_____

Notes by:	Day:	Date:	Time:

APPENDIX 4.1B: ANIMAL HAIR
EXAMINATION PROFORMA

NATIONAL CENTRE FOR FORENSIC STUDIES NCFS UNIVERSITY OF CANBERRA		Page	of			
ANIMAL HAIR EXAMINATION RECORD—Sheet 2			Case Reference:			
Hair number		1	2	3	4	5
Microscopic Feature		R➜T	R➜T	R➜T	R➜T	R➜T
Pigment density	None					
	Light					
	Medium					
	Heavy					
Pigment distribution	Towards centre					
	Even					
	Towards cuticle					
Medulla distribution	None					
	Fragmented					
	Interrupted					
	Continuous					
Medulla type	Amorphous					
	Ladder					
	Lattice					
	Aeriform lattice					
Scale edges	Smooth					
	Crenate					
	Rippled					
	Scalloped					
Distance between scales	Close					
	Near					
	Distant					
Scale pattern	Mosaic					
	Simple wave					
	Inter. Wave					
	Waved mosaic					
	Single chevron					
	Double chevron					
	Coarse pectinate					
	Lanc. pectinate					
	Irregular petal					
	Diamond petal					
Ovoid bodies						
Other						
Examined by:						
Notes by:	Day:		Date:		Time:	

APPENDIX 4.2A: HUMAN HAIR
EXAMINATION PROFORMA, LPM

NCFS NATIONAL CENTRE FOR FORENSIC STUDIES UNIVERSITY OF CANBERRA			Page		of	
HAIR EXAMINATION RECORD - Sheet 1			Case	Reference:		
Hair Number		1	2	3	4	5
Macroscopic Feature		R→T	R→T	R→T	R→T	R→T
Length	cm					
Shaft profile	Curved/Straight Wavy Curly Peppercorn					
Colour[1]	Colourless Yellow Brown Reddish Greyish Black Opaque					
Root[2]	Not present Anagen Catagen Telogen					
Tip	Distinct taper Cut Rounded Frayed or Abraded Split Crushed or Broken Singed					
General description and comments:						

[1]A assessed with stereo microscope using standardised illumination
B basic colour to be qualified by shade of depth of colour—Light (L), Mid (M) or Dark (D)
C note artificial colouring
[2]where absent, describe appearance

Examined by:_____

Notes by:	Day:		Date:	Time:

APPENDIX 4.2B: HUMAN HAIR
EXAMINATION PROFORMA, TLM

NCFS **NATIONAL CENTRE FOR FORENSIC STUDIES UNIVERSITY OF CANBERRA**		Page	of			
EXAMINATION RECORD - Sheet 2		Case Reference:				
Hair Number		1	2	3	4	5
Microscopic Feature		R➔T	R➔T	R➔T	R➔T	R➔T
Shaft diameter max units ×40						
Pigment density	None Light Medium Heavy Opaque					
Pigment distribution (across hair shaft)	Uniform Towards medulla Towards cuticle To one side					
Pigment aggregate shape	Streaked Clumped oval Clumped round					
Pigment granule shape	Fine Oval/oblong					
Pigment aggregate size	Fine Medium Coarse					
Pigment granule size	Fine Medium Coarse					
Ovoid bodies*						
Medulla distribution	None Medulla < space Medulla > space Continuous					
Medulla type	Opaque Translucent					
Cortical fusi*						
Cortical texture	Not visible or smooth Visible or coarse					
Cuticle	Thickness units ×40 Colour					
Cuticle outer	Smooth Serrated Ragged Cracked Looped					
*where present their shape, size and distribution both along and across the hair shaft may have some value when used as comparative features						
Examined by:						
Notes by:	Day:		Date:		Time:	

APPENDIX 4.2C: HUMAN HAIR EXAMINATION PROFORMA COMPARISON

NATIONAL CENTRE FOR FORENSIC STUDIES NCFS UNIVERSITY OF CANBERRA		Page of		
HAIR COMPARISON EXAMINATION RECORD		Case Reference:		
Hair Number				COMMENTS:
		Reference	Questioned	
Microscopic Feature		R➜T	R➜T	
Shaft diameter max units ×40				
Pigment density	None Light Medium Heavy Opaque			
Pigment distribution (across hair shaft)	Uniform Towards medulla Towards cuticle To one side			
Pigment aggregate shape	Streaked Clumped oval Clumped round			
Pigment granule shape	Fine Oval/oblong			
Pigment aggregate size	Fine Medium Coarse			
Pigment granule size	Fine Medium Coarse			Tech review comments:
Ovoid bodies				
Medulla distribution	None Medulla < space Medulla > space Continuous			
Medulla type	Opaque Translucent			
Cortical fusi				
Cortical texture	Not visible or smooth Visible or coarse			Conclusion:
Cuticle	Thickness units ×40 Colour			
Cuticle outer	Smooth Serrated Ragged Cracked Looped			
Examined by:				
Notes by:	Day:	Date:		Time:
Reviewer (if required)

Name: | Day: | Date: | | Signature |

APPENDIX 4.3: FEATURES OF ANIMAL HAIRS

Features that should be assessed for animal hairs:

1. **Profile** (general shape)
   Shield, straight, symmetrically thickened or wavy.
2. **Cuticle or scale features**
   - *Scale margin*: Smooth; crenate (sharp pointed teeth); rippled (indentations deeper than crenate or rounded); or scalloped (margins with broad rounded teeth).
   - *Distance between scales*: Close; near; or distant.
   - *Scale pattern*: Mosaic (either regular or irregular); wave (simple regular, interrupted regular, streaked, irregular waved, mosaic or regular waved mosaic); chevron (single or double); pectinate (coarse or lanceolate); or petal (irregular or diamond).
3. **Medulla**: Note whether present of absent. Where present, it may be *continuous, interrupted* or *fragmented*.
   In non-human hairs, it is often continuous with a defined structure. The structure can be of two main classes, *ladder* or *lattice*.
   - A ladder medulla is so called because it looks like the rungs of a ladder. Where there is a single row of "rungs", this is a *uniseriate ladder*; with several rows, a *multiseriate ladder*.
   - A lattice medulla is so called because it has the appearance of a lattice made up of "struts" of keratin which outline the polyhedral-shaped spaces, each of which is continuous with its neighbours.
   - A special type of lattice medulla, an *aeriform lattice*, differs in that the shapes giving the appearance of the lattice have arisen from cell collapse leaving air-filled gaps which are roughly polyhedral in shape.
4. **Colour**: The colour of the hair results from pigment particles deposited in the cortex. Overall, visual and macroscopic colour are important in non-human hair identification, with the detail of pigmentation in the cortex being less important that in human hair. Pigment should be assessed with respect to: amount (sparse or dense); and distribution (along the shaft (even) and across the shaft (denser near the centre; denser near cuticle).
5. **Cross-section**: Sectioning is not always carried out because it is destructive. Information that can be gained from cross-sections is three-fold.
   - Good appreciation of pigment distribution across the shaft.
   - The position of the medulla, which can be in the middle (centric) or off to one side (eccentric).
   - The shape of the hair.

However, these features can also be assessed by optical sectioning when hairs are viewed in the longitudinal plane.

APPENDIX 4.4: NUCLEAR STAINING
OF TELOGEN ROOTS

The application of nuclear specific stains to the telogen root can indicate the presence of nucleated cells retained within the root or follicular material adhering to the telogen root. Bourguignon *et al* (2008) and Boonen *et al* (2008) showed that nuclear specific staining of telogen hair roots resulted in a determination of the number of nuclei still present in or around the telogen root by fluorescence microscopy. They visualised nuclei in telogen hair roots with DAPI (40–6 diamidino-2-phenylindole), a blue fluorescence stain excited at the ultraviolet (UV) wavelength, Bourguignon *et al* (2008). To view and count the stained nuclei in the telogen roots they used a transmitted fluorescence light microscope equipped with a DAPI specific filter Boonen *et al* (2008). Bourguignon *et al* (2008) and Boonen *et al* (2008) achieved a 79% success rate of complete and incomplete STR profiles from telogen hairs with nuclear counts >20 cells and for telogen roots with >50 counted nuclei the success rate increased to 88% successful STR profiles. To extract the nu-DNA, Bourguignon *et al* (2008) and Boonen *et al* (2008) followed the protocol of Hellman *et al* (2001) where the hair root was cut from the hair, digested and organically extracted.

Harris's haematoxylin is an alternative nuclear stain that does not require a fluorescence capable microscope. Harris's haematoxylin is widely used as a nuclear stain in the fields of histology, pathology and cell biology (Spector and Goldman, 2008) where the nuclear material of the cell is stained purple and can be visualised using bright field microscopy. Based on this recent research and the validation study undertaken (*Boonen et al*, 2008; Bourguignon *et al*, 2008; Brooks *et al*, 2010), it appears that when such nucleated cells are stained, counted and extracted for nu-DNA then a minimum of 30 cells yields a reportable DNA profile. Current nu-DNA extraction protocols, however, may be more efficient and produce reportable STR profiles with fewer nucleated cells.

1. **Reagents required:**
   - Absolute ethanol (Sigma)
   - 20 mL DAPI (40–6 diamidino-2-phenylindole)
   - DABCO (1′-4-diazabicyclo (2.2.2) octane)5,6 - 2.24 g/10 mL
   - 0.2 M Tris HCl pH 7.4
   - Harris's haematoxylin stain (pre-made)
   - 100% glycerol
2. **Workplace health and safety**
   General workplace health and safety (WHS) requirements during the nuclear staining procedure of stage 2 and stage 3 telogen root types refer to Work Health and Safety regulations.

- Harris's haematoxylin—Nuclear Stain—Irritant
- Absolute ethanol—Flammable, respiratory tract Irritant
- DAPI—Harmful if swallowed or absorbed through skin
- DABCO—Harmful if swallowed, very destructive to mucus membranes
- Tris HCl—Irritant.
- Glycerol—Very mild irritant.

3. Procedure

3.1 Apoptosis in Hair

Biologically, the telogen hair root is in an apoptotic and keratinisation stage, thus generally containing limited and often degraded nu-DNA (Glucksmann, 1951; Wyllie, 1980). A simple fixation step helps to reduce the loss of already degraded nu-DNA that is no longer encapsulated within the nucleus. Chemical fixation is routinely used for biological tissues because it terminates any ongoing biochemical reactions, such as those found in the biological functions of apoptosis and necrosis (Brooks *et al*, 2010; Glucksmann, 1951; Wyllie, 1980). Fixation employs agents that permeate tissues and cells and combine covalently with their major biochemical constituents (lipids, proteins and carbohydrates) to "fix" them within the tissue (Brooks *et al*, 2010).

3.2 Selection of Hairs for Nuclear Staining

Hairs in the telogen root stage require separation into the 3 stages identified and illustrated in Figure A4.1—unstained telogen roots. Stage-1 telogen hair roots (A) can be returned to the collection "paper boat", sealed and stored with the item for further examination by comparison microscopy as required. Telogen hair roots exhibiting stages 2 (B) and 3 (C) are then prepared for fixation and nuclear staining (Spector and Goldman, 2008; Brooks *et al*, 2010).

> *Note: If a number of hairs with telogen roots at stages 2 and 3 are recovered from the same item these hairs can be processed together in the fixation and staining protocols. When the nuclear counts have been completed then the individual hairs should be itemised if required.*

Note for Figure Chapter 4, Appendix 4.1:

Type 1 Unstained (a): The majority of telogen roots observed will look like this hair root. Club or cotton bud shaped (chevron), no remnant sheath material and consequently not useful for either nuclear staining or DNA analysis. Both the DAPI- and haematoxylin-stained type 1 roots have no nuclei stained confirming the initial assessment.

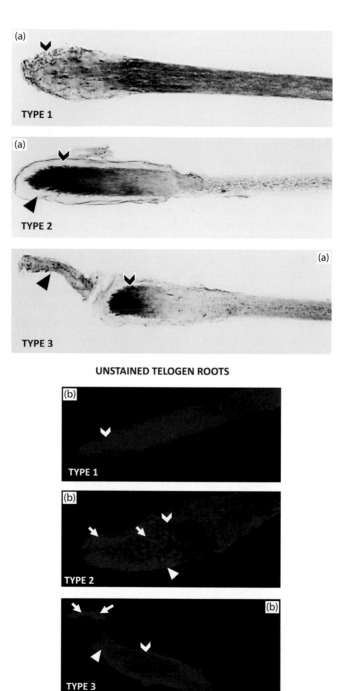

FIGURE A4.1   Different telogen hair root types with nuclear staining. (a) Unstained telogen roots. (b) DAPI-stained telogen roots. (*Continued*)

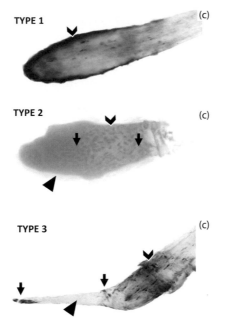

TYPE 1   (c)

TYPE 2   (c)

TYPE 3   (c)

HAEMATOXYLIN STAINED TELOGEN ROOTS

FIGURE A4.1 (*Continued*)   (c) Haematoxylin-stained telogen roots.

Type 2 Unstained (a): Observed clearly is the telogen shaped root (chevron) but here it still has some sheath material attached (arrowhead). This is a good candidate for nuclear staining with either DAPI or haematoxylin, as shown in the two stained examples. The small bright blue or purple spots (stain dependent) indicate an individual nucleus (arrows) that can be counted and subsequently analysed for nu-DNA.

Type 3 Unstained (b): Type 3 roots are also obviously at the telogen stage of growth (chevron) but with much less remanent sheath material (arrowhead) including a follicular tag still attached. Type 3 roots may be a reasonable source of nuclei but sometimes yield no nuclei post staining. As shown in both the DAPI- and haematoxylin-stained Type 3 roots the majority of the nuclei are located in the "tag" end (arrows). Depending on the number of nuclei counted (<30) will determine the decision for nu-DNA analysis.

3.3 Fixation, Visualisation and Nuclei Counting of Telogen Roots

The process of visualising and counting the nuclei in the telogen hair roots is a simple and rapid five-step process. Once the hair is identified as having a suitable telogen root for staining, the fixation, staining, mounting and counting of nuclei takes no more than an hour.

3.3.1 Fixation

- Hairs remained whole—do **NOT** cut off the root end.
- All hair roots are chemically "fixed" in absolute ethanol for 30 minutes.

3.3.2 Staining

i. *DAPI-DABCO Staining:*
   - Post fixation, transfer hairs to a glass tube containing the DAPI staining reagent comprised of 20 mL DAPI, 400 mL DABCO (2.24 g/10 mL 0.2M Tris HCl pH 7.4) and 400 mL 100% glycerol (Brooks *et al*, 2010).
   - Wrap tube in an aluminium foil and leave in the dark to stain for 5 minutes (DAPI is light sensitive).
   - Remove the stained hairs from the DAPI staining reagent and place on a filter paper.
   - Immediately mount individual hairs in glycerol (one drop of glycerol will be sufficient) on a glass microscope slide (1× hair per slide) and place coverslip on slide and seal with clear nail varnish.
   - If slides are not viewed immediately, cover with foil to keep dark and place them in a refrigerator.

ii. *Harris's Haematoxylin Staining:*
   - Post fixation, transfer hairs to a glass tube containing 1 mL filtered Harris's haematoxylin, and leave for 5 minutes to stain (Brooks *et al*, 2010).
   - Rinse the stained hairs in distilled water, and mount (1× hair per slide) as described for the DAPI-stained hairs.

3.3.3 Counting Stained Nuclei

- Using a compound light microscope in transmission mode, examine the DAPI* and haematoxylin** stained hairs, and count the number of nuclei present.
- The nuclei in the DAPI-stained hairs will appear as small blue spots—see Figures A4.1 (b) and A4.2— DAPI-stained telogen roots.

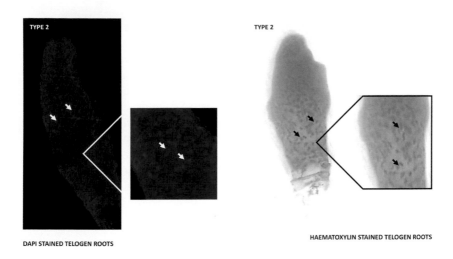

FIGURE A4.2 Stained nuclei remnant sheath material. Arrowheads indicate individual stained (DAPI and haematoxylin) nuclei still viable within the remnant sheath material attached to both these type 2 telogen roots.

- The nuclei in the haematoxylin-stained hairs will appear as small purple spots—see Figures A4.1 (c) and A4.2—haematoxylin-stained telogen roots.
- Count visible cell nuclei and record the hair root type and number of nuclei in the notes.

4. **Post-Nuclear Counting**
- Identify any telogen roots that have a count of 30 or more nuclei.
- Remove the coverslips from the microscope slides by running a sterile scalpel blade along the edges of the coverslip and lifting it off.
- Remove the individual hairs from the slide and place in separate sterilised, labelled 1.5-mL Eppendorf™ tubes.
- Rinse hairs with less than 30 nuclei in distilled water, dry on filter paper and place in a paper boat with the other recovered hairs, then return to original item packaging, for further microscopy examination if required.

5. **DNA Analysis**
- Stained telogen hair samples containing >30 nuclei should be extracted and analysed for nu-DNA.

---

* **Note:** If using DAPI staining a microscope equipped with a fluorescence capability is required using the UV filter cube.
** **Note:** When viewing haematoxylin-stained hairs, roots can be observed using Bright-Field illumination on a compound microscope in the transmission mode.

APPENDIX 4.5: TEST FOR BLEACHED HAIR

Cosmetic hair treatments are common. However, in forensic hair examination it is unusual to comment beyond noting that the hair has undergone artificial colouring. Some hair treatments such as bleaching, permanent waving, relaxing and permanent dyes have an oxidative effect on the hair. Chao (Chao *et al*, 1979) reported that such treatments alters, to some extent, the oxidative cleavage of the disulphide bonds that can be detected by several methods including FTIR, electron spectroscopy and methylene blue staining. The simplest and fastest method by far is the latter one.

One case known to us required determining if the victim's hair had been artificially coloured. We used the method described by Roe (Roe *et al*, 1985) who reported successfully detecting oxidative hair treatments with methylene blue. Their method can be summarised stepwise as follows:

Method for Methylene Blue Staining

1. Make up an aqueous 0.5% solution of methylene blue (basic stain).
2. Completely immerse hair/s in this solution for 4 minutes.
3. Remove methylene blue solution and rinse twice in distilled water.
4. Mount hairs in preferred mountant and examine by LPM or TLM.
5. Methylene blue has an affinity for the electron dense sulphonic acid and is taken up by the hair in the regions of oxidative treatment turning these regions bright blue (see Figure 4.7).
6. Observation by microscopy shows the blue delineations of the effected hairs/or regions of the hair, indicating oxidative treatment.

CHAPTER 5

Evaluation and Interpretation

5.1 INTRODUCTION

In Chapters 1–4, we discussed the potential value of forensic examination of human hairs; outlined the basic biology of the formation and growth of hairs that underpins forensic application; emphasised the essential role of a crime scene examiner in recognition, recording and recovery of hairs and detailed the microscopic features of hairs that can be observed and assessed against a four-level approach to hair examination.

In this chapter, we will discuss the evaluation and interpretation of information gained from the examination of hairs against a framework of "fit for purpose" outcomes for investigative purposes and for potential use as evidence in courts of law. We will draw on AS 5388.3-2013 Forensic Analysis Part 3: Interpretation, to guide our consideration of this topic.

5.2 TRANSFORMING DATA INTO INFORMATION

For those of us who really like looking at hairs, the microscopic examination of a hair is exciting as the detailed features are observed and a full picture of the overall pattern of features is revealed! However, we recognise that not everyone will automatically share our enthusiasm for hair examination. This has potential consequences. We have previously commented on the fact that less attention will be paid to the potential value of hairs as evidence where the person tasked with the recovery of such evidence does not hold a positive view of the value of examining hairs. This flows on to the microscopic examination process. A lack of enthusiasm, either at the level of the organisation or the individual, is unlikely to see any degree of commitment to, or investment in, hair examination. It does not follow that individual enthusiasm translates

DOI: 10.4324/9781315210650-5

into high performance. Hair examination is essentially an observation science, and not all individuals will have equal abilities when it comes to microscopic examination. However, we do know that a good starting point is interest and enthusiasm.

Drawing from the world of medical radiology, it has been shown that a person interested and enthusiastic about seeing will see more than the one who is not so inclined (Tuddenham, 1962).

In another study involving radiologists, Thomas (1969) found that, when looking for pathological features on an X-ray film, instead of examining the entire film each radiologist had a distinctive pattern of eye search movements, and that all examiners did not look at the same areas of the film in reaching a conclusion. He also found that students fell into two main groups, a small group that was able to see quickly and effortlessly and a larger group that was only able to see slowly and with difficulty. Gaudette draws a comparison between this process and that of hair examination, suggesting that examiners need to develop a "creative visual gestalt" introducing the concept of *visual literacy*, as the hair comparison process is more of a pattern recognition process than a logical step-by-step one (Gaudette, 1999). However, this is not to say that records of detailed features should not be recorded as this step-by-step process forces the examiner to look more carefully and in turn this improves visual literacy (Gaudette, 1999).

Gaudette also argues that there is no substitute for gaining experience through conducting many hair examinations and comparisons to develop visual literacy whilst recognising that different individuals will have different levels of discrimination. Hence, training, experience, general competency and appropriate quality checks are critical components in ensuring high standards (Gaudette, 1999). Koch *et al* (2020) also stress the importance of visual acuity, training and experience of individual analysts as the foundation for microscopical analysis and on the ability to process pattern differences within the human brain.

However, it has been argued by some that experience is not to be trusted and one must have some empathy with that view considering cases such as the previously discussed case of Driskell (Anon, 2006).

AS 53388.3-2013 Forensic Analysis Part 3: Interpretation gives us a way forward where experience is assessed in a broader context of *professional judgement*. Appendix B of AS 5388.3 (modified here as Figure 5.1) gives a visual representation of the various factors that need to be considered in informing professional judgement.

In Figure 5.1 the vertical decision pathway starts by posing the *relevant* question. We will return to this key element as this is often where the problem starts in the forensic process—put simply, if you do not ask the right question, then you cannot get the right answer.

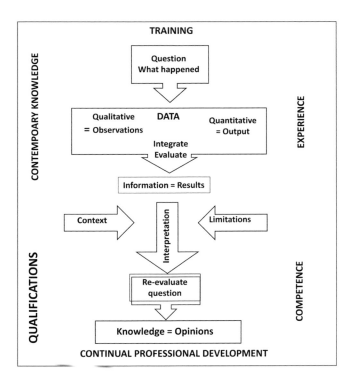

FIGURE 5.1 Visual representation of the various factors needed in informing professional judgement.

In a scientific process, this question would be framed as the null hypothesis and its alternative—these can also be stated as so-called propositions. AS 5388.3 defines the null hypotheses as follows:

NULL HYPOTHESIS

1. When comparing two objects or sets of statistical data, the hypothesis that states that an observed difference (as between the means of two samples) is due to chance alone.
2. The general or default position. For example, the null hypothesis might be that there is no relationship between the two items.

The *alternative hypothesis* is that there is a statistically significant relationship between two items.

We would argue that the starting point for the hair examiner should always be that there is no relationship between hairs being examined and compared. To show there is no relationship, the mindset must be to look for differences and NOT similarities. In our view, it is unfortunate that the term "similar" continues to be used as it gives no real sense of discrimination. In principle, one should be able to show two items are different, whilst it is impossible to prove that two items are identical.

Following the scientific method, challenging the null hypothesis requires data to have been collected. This can be *qualitative* and *quantitative* data. The microscopic examination of hairs obviously will largely generate *observations* or *qualitative* data. As we have mentioned numerous times, the key challenge is to try and capture these observations through a well-structured, systematic and thorough process that maximises the accuracy and repeatability of these observations. Regardless of the level of hair examination, the forensic scientist will then evaluate the data and make decisions in the context of the specific case circumstances. This is the process of evaluating the data and turning it into information. At this point, the scientist should consider any limitations and contextual information before reaching an opinion. It may be necessary to reset or re-evaluate the question at this stage in the process. AS 5388.3 defines contextual information as follows:

Contextual Information

Another factor to be considered while evaluating data or interpreting information is the context of the information.

Although knowledge of the case could increase the risk of bias, relevant contextual information might be necessary for the derivation of knowledge. A context could be required to properly evaluate competing questions or might well affect how well information fits expectations from the circumstances of the case. In these circumstances, however, steps need to be taken to counter any contextual bias that might affect evaluation or interpretation.

In the real world what this simply means is that you have a responsibility to ensure that your examination attempts to answer relevant questions with *useful* knowledge. Useful knowledge is defined as a collection of information intended to be useful to a particular context—this may seem self-evident, but it is important to not lose sight of the fact that the primary objective of a forensic examination is not the search for knowledge in its own right but rather the search for knowledge that is informative in the context of an investigation and/or for the purposes of forming an opinion. Forming that opinion requires the use of *professional judgement*. This is defined in AS 5388.3 as follows:

Professional Judgement

Forensic science requires professional judgement. Professional judgement is influenced by qualifications, training and experience. The value of professional judgement is determined by competence, knowledge and professional development.

Professional judgement can also involve consultation with peers and consideration of their input.

The key element of a professional judgement is that the individual has appropriate qualifications and training fit for the purpose. As there is no specific qualification to be a hair examiner, the general requirement is that the scientist must have relevant academic qualifications, most often a science degree, general training in forensic requirements and specific training in hair examination relevant to the level of hair examination in the scope of a laboratory.

In our view, the value that should be placed on the findings of any individual examiner should then depend on them demonstrating they have a level of specialist knowledge gained through study, ongoing training and professional development to maintain *contemporary knowledge* appropriate to their claimed expertise.

Experience can be useful, but this needs to be experience relevant to the speciality and where the examiner has also demonstrated *ongoing competence* through appropriate proficiency testing.

AS 5388.3 requires that there is a formal process of review of information to ensure the accuracy of data, the correctness of transcripts and the appropriateness of any evaluation process. Such a review should be conducted by a second qualified practitioner. AS 5388.3 requires that the review procedures are documented and that they specify the following:

1. The proportion and type of examinations reviewed.
2. How and when a review is conducted and recorded.
3. The principal responsibility for the information is retained by the primary practitioner.

5.3 FORMULATING AN OPINION

Assuming the individual can meet the above requirements, understands the limitation of their expertise and does not claim greater levels of expertise than can be justified, they are entitled to offer an opinion. Although the *opinion must be owned by the individual*, this does not preclude the individual from consulting with peers during any stage of the process and considering their input as outlined in the standard. AS5388.3 states that the process of formulating an opinion should include the following:

1. Considering relevant questions and their related explanations.
2. Evaluating the explanations according to their ability to explain the information.
3. Ranking conclusions according to the explanations and support provided by the information.
4. Giving an opinion as to the best supported conclusion.

The standard also states that any *identifiable and reasonable* alternatives to the final opinion should be documented as should the reasons for their rejection. Where there is more than one explanation that could explain the data, and none of the explanations can be rejected, the explanations should be ranked and the reasons for each ranking included in the case notes.

The underlying purpose is to ensure that the basis for all opinions is clear to all similarly trained and experienced practitioners—peers—but the standard goes further in that it also requires that the opinion, and the reasoning behind the formulation of that opinion, should be expressed in clear and simple terms that can be understood by a non-expert.

5.4 ESTIMATING PROBABILITIES

The standard also considers the weighting of opinions expressed in numerical terms as a probability or ratio of probabilities in a way that demonstrates the practitioner's certainty of that opinion. With respect to hairs, no formal database exists that can be considered as being representative and statistically valid from which reproducible numerical values can be obtained that would form the basis for calculating a *likelihood ratio* or producing *frequency data* to support a level of discrimination.

AS 5388.3 states that, *in the absence of a formal database*, practitioners may estimate the frequency of occurrence of an evidence type based on their experience. The standard goes on to state that this may be expressed as part of a likelihood ratio and used in a qualitative way or as a frequency statement. The statement represents the level of support for the conclusion expressed in words, rather than numerically, as to the significance of the results obtained. The "words" used should provide a clear indication of the level of support for the conclusion or opinion. We will return to this topic in Chapter 6 on reporting.

The standard makes it clear that the approach used by the practitioner *shall be* stated in their report—not optional—and that the approach used must be applicable to the type of examination and may depend on supporting data.

It is important to point out that AS 5388.3 was developed to be a general standard with the intention that each discipline would then develop its supporting standard, detailing how the overall standard should be applied in practice to their discipline. Regrettably, no such discipline-specific standard has been developed for hair examination.

5.5 DEFINING ERROR

It is important to recognise that all techniques have inherent limitations and the examination of hairs is no exception. Where data is numerical and quantitative, the accepted practice is to consider *uncertainty of measurement* which can give a statistical weighting in interpreting results. In qualitative examinations, such as is the case with hairs, variation exists, but it is not normally suitable for the kind of quantitative analysis that yields an uncertainty value. In the context of scientific use measurement uncertainty can also be termed *statistical error.*

Statistical error is the difference between an observed or calculated value and a true value.

It is not possible to calculate statistical error for microscopic observations from hair examinations. Hence, we must default to think of error as meaning a mistake or an incorrect result. These are usually defined as *type 1 and type 2 errors.*

AS 5388.3 defines type 1 and type 2 errors as follows:

- *A type 1 error* is where the test incorrectly rejects the null hypotheses when the null hypothesis is true. A type 1 error is often referred as false positive.
- *A type 2 error* is where a test incorrectly supports the null hypothesis when the null hypothesis is in fact false. A type 2 error is often referred to as false negative.

Figure 5.2 shows a diagrammatic representation of how type 1 and type 2 errors are related. It follows that decreasing type 1 errors (false inclusions) increases type 2 errors (false exclusion) and vice versa.

In the context of a hair examination, the null hypothesis is that there is *no relationship between two samples.* If the examiner concludes that there is no relationship between a recovered hair and a known sample, then the

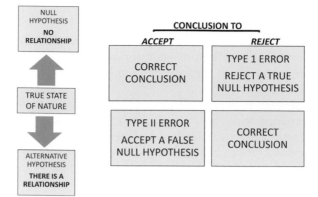

FIGURE 5.2   Relationship between null and alternative hypotheses and type 1 and type 2 errors.

recovered hair is correctly excluded. If the examiner concludes that there is a relationship, then this is a false inclusion or type 1 error.

The alternative hypothesis is that there is *a relationship between two samples*. If the examiner concludes that there is a relationship between a recovered hair and a known sample, then the recovered hair is correctly included. If the examiner concludes that there is no relationship, then there is a false exclusion or a type 2 error.

As we have already stated, the appropriate null hypothesis, or starting point for a hair examination, is that there is no relationship between a recovered hair and a known sample. It is impossible to prove a relationship, but it is possible to disprove a relationship through looking for meaningful differences. A question then arises, whether a type 1 error is more serious or a type 2 error? In our view, this is where some of the criticisms of hair examinations fail to understand the purpose of hair examination. The usual starting point for commentators is that the primary purpose of a forensic examination of hairs is to individualise to a point where some claim is made of uniqueness. The latter is simply not possible and certainly cannot be proved in any scientific way. On the other hand, it can be shown that hairs could not have a common origin based on differences.

Hence, in order to evaluate the potential problems that can arise from type 1 and type 2 errors, it is essential to be clear about the purpose of the examination in the context of being *fit for purpose* using our four-level approach to hair examination.

> *Level 1 examinations* deal with the recognition and separation of human and animal hairs. Hence, the null hypothesis would be "the hair being examined is not a human hair" with the alternative hypothesis being "the hair being examined is a human hair". If the *ground truth* is that the hair is an animal hair, then a type 1

error would be where the examiner identified the hair as being of human origin or a false inclusion. This would have some practical consequences and could lead to either no further work being conducted or to unnecessary work being conducted, not to mention it being somewhat embarrassing if the hair was sent to a level-4 expert for it to be shown to be a type 1 error! From our personal experience this type of error does occur in practice. Therefore, it is important that, even at this level, examiners have received appropriate training and have demonstrated competence. In real life there will be cases where the examiner may be unsure about making a definitive call. In such circumstances it is appropriate to have an **inconclusive opinion** provided it is reported with an explanation. For example, there may only be a fragment of hair with insufficient microscopic details to reach a definitive decision.

Level 2 examinations deal with the examination of human hairs and selection for DNA analysis. Initial examinations should focus on selection of hairs for nuclear DNA (nu-DNA) analysis based on classification of the growth stage of the hair root where a root is present. Whilst we prefer to examine hair roots with mounted hairs (and of course this must be the case for hairs stained for nuclei), we recognise that in many organisations hairs are examined unmounted, either with the naked eye or with low-power microscopic examination (LPM). With a null hypothesis that the hairs are not suitable for DNA analysis a type 1 error would be to select hairs that are suitable for analysis. However, as this selection is a presumptive level test, false positives are preferable to false exclusions where potential evidence may be lost. In this instance, it is preferable to select all hairs that are *non-telogen* and, if in doubt, include and do not exclude. This should not be an excuse to include everything as this would defeat the purpose of the process, that is, to minimise the number of unsuitable hairs going through for DNA analysis. In laboratories that include a further test for the presence of nuclei, the laboratory procedure should be clear as to the acceptance criteria for non-routine nu-DNA testing.

Other microscopic examinations at this level include determination of somatic origin and ethnic origin. Whilst these determinations may have case significance, the consequences of a type 1 or 2 error are relatively minimal. Notwithstanding, examiners at this level should again be able to demonstrate competence through appropriate training and proficiency testing. Body area determinations and, especially ethnic determinations, are sometimes not clear and opinions may be inconclusive or qualified.

Level 3 examinations deal with detailed microscopic examinations and comparison microscopy. Initial microscopic examinations will be at LPM and usually involve either capturing an initial description of the microscopic appearance of a hair or hairs using a stereo microscope with up to ×40 or so magnification. Where there are one or more known hair samples from complainants and suspects, along with one or more recovered hairs, the forensic process will be to examine the recovered hairs and to determine whether they are different from any of the known samples. The null hypothesis is that the hairs being "compared" (not at this stage actual comparison microscopy) are not from a common origin, i.e., no relationship. If the recovered hair is found to be different then it will be excluded. If the examiner gets this wrong, then the hair will be wrongly excluded, and potential evidence will be lost—false exclusion or a type 2 error. If the examiner forms the opinion that the hairs cannot be excluded but is in fact wrong, then this is a false positive or type 1 error. The question is whether at this level of examination the consequences of a type 1 or type 2 error is greater? If one views this level of microscopy as a "confirmatory test" then the aim should be to minimise the number of false inclusions or positives, i.e., type 1 errors. However, as we have seen, minimising the number of type 1 errors will increase the number of type 2 errors or false exclusions. The key at this level of examination is to recognise that this is just a confirmatory test to assist in selecting those hairs that may warrant further examination. Hence, **in this context,** false exclusions are a bigger problem than false inclusions. It should go without saying, but needs to be said, that the easy way to avoid false exclusions is of course to exclude no hairs!! This would not be helpful and is to be actively discouraged. The purpose of selection of hairs at this level is about case triage informing the selection of hairs for either more detailed microscopy and/or mitochondrial DNA (mt-DNA) testing.

For mt-DNA testing this stage will be a failure, and not fit for purpose, if subsequent analysis shows that numerous selected hairs are not from the proposed source. However, equally it should be expected that some selected hairs may not be from the proposed source. From this perspective we would argue that if in doubt (within reason!) then leave it in rather than leave it out! Finally, not all hairs have numerous features which offer discrimination potential and, hence, hairs may go forward to mt-DNA testing simply because there is no potential for further meaningful microscopic examination. It is wrong to classify these as an error as identification is not the aim of this level of examination.

For many, if not most laboratories, LPM examination of hairs is as far as the examination process goes.

For some laboratories recovered hairs, that have not been excluded, will then be subject to examination with transmitted light microscopy (TLM) at magnifications of up to ×400 (objective ×40). As this is undoubtedly a time-consuming process, it should only be undertaken in laboratories where there are scientists who have received training at this level of examination and who have proven competence through appropriate proficiency testing. The latter training and testing are not simple or easy to design or conduct. We will deal with training and proficiency testing in Chapter 7. It is easy to design and set up very simple tests, but these will not test the genuine ability of the examiner to differentiate based on their *visual literacy*. It is possible to train any individual to recognise many of the microscopically observable features seen in hairs; for example, ovoid bodies or cortical fusi. It is much more difficult to train individuals to reproducibly classify pigment features and patterns and one must accept that there is some inherent ability in individuals to see patterns—this is not unique to hairs and applies across many feature-based disciplines. However, even with an inbuilt ability to see patterns, there is also a degree of "practice makes perfect". This is probably not the best way to express this, but anyone who has seriously spent time looking at hairs knows that if there is a significant time gap when they have not examined hairs, it takes time to get "their eye in". For this reason, even experienced hair examiners can lose confidence—hair examination is not an occasional pastime! In our view, unless a laboratory is willing to support a reasonable volume of hair casework at this TLM level of examination, then *current competence* is impossible to maintain.

On the assumption that the examiner is currently competent, a detailed description of *relevant* recovered hairs must be recorded using a checklist to ensure a systemic and thorough process. The checklist should be enhanced by appropriate descriptions and images where useful to record the presence of features. Through this process the examiner will capture a detailed description of not only the individual features but how they are displayed along the length of the hair shaft. Case circumstances will determine if every recovered hair needs to be examined or a selection of recovered hairs. The same process needs to be repeated with a representative selection of hairs for one or more known hair samples. At this point, the examiner will be in a position to choose known hairs to directly compare with a recovered hair. In order to not exclude a known sample as a possible source of a recovered hair, there must be no meaningful differences between hairs being compared. This is not to say that any two hairs will be identical, but any unexplained differences should result in the known sample being excluded as a possible source for a recovered hair. Even in a relatively simple case, there may be many recovered hairs and at least two known samples. In a more complex case, there may be multiple known samples and numerous recovered hairs.

Hence, it is important to try and achieve as high a level of discrimination as possible at the LPM stage as this will reduce the number of recovered hairs to be examined at the TLM level.

At the comparison stage, the primary purpose of the examination is still to exclude based on observed differences. If a known sample cannot be excluded as a possible source for a recovered hair or hairs, then we move from exclusion to potential inclusion. If the purpose is to select hairs for mt-DNA testing, then the primary role of this much more detailed examination will still have been to exclude hairs and select hairs that are most likely to have come from a known source. If mt-DNA testing is not available, then the examiner must offer an opinion such as "cannot exclude" or "could have come from".

There is also an obligation to try and assess the "weight and substance" of any conclusion. This is not an easy matter as it is simply not possible to place a statistical figure on the conclusion. However, in our view neither is it acceptable to agree with suggestions such as "all brown hairs are the same". Without ever crossing the line to give an impression that it is possible to identify a person based on hair microscopy, hairs from different individuals are often clearly different. How do we assess how different? One analogy might be to consider the examination of a sheet of paper. Presented with two sheets of white paper with no writing and only visual examination, discrimination may be possible based on size, any visible damage or imperfections. The hair equivalent of this scenario would be two essentially colourless hairs with no other microscopically visible features. There may be differences in the length of hairs or damage characteristics. If we now move to the two sheets of paper having a small number of words scattered across the sheets, then there is greater potential to discriminate as there is potentially more information to examine. The hair equivalent would be two hairs with visible microscopic features that can be described and compared. The greater number of words or features increases the potential to discriminate. However, if the density of words is so great, or they are obliterated and cannot be read, then although there is a large amount of potential information, it cannot be deciphered. The hair equivalent is where the density of pigmentation is so great that light cannot penetrate the hair shaft and microscopically it appears opaque. For example, this is often the situation with hairs of Asian origin.

Hence, **the way to think about a hair is that it is a package of potential information**. The role of the hair examiner is to interpret the information in the context of the case in question. This is not only about selecting hairs for DNA analysis or reaching some degree of individualisation. As we have discussed in earlier chapters, some of the potential information is of a criminalistics nature, can answer investigator questions and can assist in answering the what happened questions often of most interest in court proceedings.

CHAPTER 6

Reporting

6.1 SCOPE

In this chapter, we move from the evaluation and interpretation of hair data to the reporting of results and opinions. We will draw as relevant on Australian Standard AS 5388.4-2013.

Depending on the investigation, or enquiry, one or multiple reports may be provided. A report may be *oral* or in *writing*. Increasingly written reports will be provided in electronic format and not in hard copy. Whether in written or in electronic format, all reports should be reviewed prior to issue and, where this is not possible, the communication should include a caveat to this effect.

Whilst recognising the importance of early and ongoing communication of results, and that information can be shared before a final report is issued, this places a heavy responsibility on the author of a report (oral or written) to ensure they accurately record such communications and ensure that they meet the same standards as would be expected of a finalised report. Where e-mail communications are used these should be treated as a report and are subject to the same standards as a written report.

In general, reports should be fit for the intended purpose, be readily understood by the intended audience, state what was done, what was concluded and any limitations with the process that would affect the conclusions.

As stated in Chapter 5, terminology is important in ensuring that reports are understood by non-specialists. This may mean that a report needs to be accompanied by a glossary of terms where the specialist use may differ from common use or understanding. All communication should be as concise as possible and at the same time unambiguous.

Finally, all communications are open to discovery and should be written in a way that reflects the requirements of relevant codes of conduct as well as any legal or organisational requirements.

DOI: 10.4324/9781315210650-6

6.2 REPORT FORMATS

The format of a report will be determined by several factors. These include jurisdictional legal requirements, the purpose of the report and the type of information that the report contains.

Oral reports may include any communication given to an investigator, lawyer or other party over the phone, by video or by any other electronic means. In general, such communications are intended to be for information and can be at any stage in an investigation or in proceedings that follow. Where possible, oral communications should be recorded. Where the latter is not practical, a written record of the communication should form part of the case file. A follow-up e-mail to confirm that both parties agree on the outcomes of the communication is highly desirable. As this type of communication can have the potential to influence the direction of an investigation, or in the preparation for a trial, it is vital that there are no misunderstandings between parties involved.

Oral communications are important in ensuring that relevant information is shared in a timely manner, but by their nature information may be given before analysis is complete. This needs to be understood by all parties to such communications. Such communication will not generally substitute or replace a finalised written report. In circumstances where examinations have not resulted in any meaningful information an oral report may be sufficient and fit for purpose.

Written reports include e-mails and text messages, interim reports, supplementary reports, laboratory reports and a full report intended for court purposes.

The use of *text messages* and *e-mails* to communicate information is an everyday reality. It is important to remember that all such communication should not be considered as "informal" and it is open to discovery. Hence, it is just as important to ensure that messaging in these formats meets the same standards as any other form of communication.

An *interim report* is one that is provided prior to the finalisation of a full report and usually intended to deliver preliminary results or a summary of final results. Such reports are most often issued to investigators to assist with the investigative process, but in some circumstances, lawyers may ask for an interim report in preparing for a court hearing. Great care needs to be taken in formulating conclusions and opinions to be included in an interim report as these may later need to be modified or changed in the light of further analysis. All reports should include consideration of any limitations and this is no less important with an interim report.

A *supplementary report* may be issued to any other form of report where further testing or information is available.

A *laboratory report* is usually a written report provided to investigators where a full report for court purposes is not required. This may be in a circumstance where the primary purpose of such a report is for intelligence purposes or where a decision has been made not to proceed with an investigation. Regardless of the "intended" purpose this does not preclude such reports eventually being used for court purposes and, hence, although they may not fully conform with the format or content of a report intended for court purposes, they should meet all other quality requirements.

A *full report* intended for court purposes should comply with all of the quality requirements for such a report. These are discussed in detail in Section 6.4.

For some forensic applications, **certificates** may be issued. These usually deal with identification or quantity of a material being analysed such as an alleged illicit drug, an alcohol measurement or the identification of cannabis. The format and wording of these certificates is usually mandated and defined by the relevant jurisdiction. Other than drugs in hairs it would not be expected that hair findings would be dealt with in a certificate format.

6.3 ISSUING OF REPORTS

AS 5388.4 requires laboratories to have documented procedures for the issuing of all types of report. These procedures need to address the circumstances under which reports are issued, circumstances in which it may not be appropriate to issue a report and the types of report that may be issued. It should be made clear on all reports their intended purpose although, as previously discussed, this does not preclude the recipient making an alternative use of the report.

Some organisations no longer issue hard copy reports with reports either being sent as an electronic communication or, in some circumstances, being directly accessible from a website by a client. In the latter circumstances, a report should be issued in a protected format. Organisations should have policies in place, and the technical capability, to detect any unauthorised change or loss of a report. The original report should meet the required standards for digital evidence.

In some circumstances, a report may have to be given a security classification which would usually restrict to whom and how it would be issued.

We will not cover the administrative and technical review in detail. Suffice to say that **all** forms of report are subject to the requirements of administrative and technical review. This includes oral communications where procedures need to be in place to review a proportion of such reports.

6.4 REPORT CONTENTS

The precise format and content of reports are subject to jurisdictional requirements. Many jurisdictions will have formal procedures or practice guidance which must be complied with. Hence, the following is a general guide as to what would reasonably be expected in a full report for court purposes.

6.4.1 General Requirements

A written report may be expected to include the following:

1. The date of issue
2. The name of the forensic facility or organisation
3. A unique case identifier
4. The name of the person responsible for the report
5. A means for ensuring that each page is a part of the report
6. A means of signifying the end of the report

A report intended for legal proceedings should in addition include the following:

1. Information about the collection and continuity of forensic material
2. Analysis and comparisons conducted
3. Results
4. Limitations
5. Conclusions and opinions
6. Qualifications and experience of the author
7. Relevant protocols
8. Definitions or explanations for technical terms used in a report

Reports whose primary purpose is not intended for court may not include all of the above.

The order of the above may vary according to jurisdictional requirements. It can be helpful to have a summary of the main findings, conclusions and opinions at the start of the report.

In some jurisdictions reports for legal purposes will be attached to a *formal statement* that may contain qualifications and other information. Some information may also be included as appendices to the report. In some jurisdictions, the format of the report has to be as a statement. However, the usual practice is that expert reports are attached to a statement.

In Chapter 3, we covered the recognition, recording and collection of hairs at a locus or scene and during subsequent laboratory examinations. The degree of detail included in a full report about the collection/recovery of forensic material will vary depending on the complexity and size of the case. However, it should be clear to the reader of the report the reasoning why samples were collected, or were not collected, and any matters that may affect the integrity of the items for the intended examination and analysis.

It is becoming more common for organisations to have some form of centralised evidence searching and triage function where forensic materials will be recorded and recovered. This may be the point at which level 1 and 2 examinations are conducted with recovered human hairs being assessed for their suitability for routine nu-DNA analysis. Hairs that are not suitable for nu-DNA analysis may be retained and not examined unless there is the potential for further examinations to answer questions of a criminalistics nature or where nu-DNA is not successful.

It is important that the "specialist" hair examiner is aware of all relevant information relating to the recording and recovery of any hairs submitted for examination. For example, where hairs have been recovered from a scene or item there should be a sampling plan clearly stated in relevant case notes. The organisation responsible for crime scene work, and any evidence collection and triage function, should have appropriate policies and methods in place for the recording and recovery of hairs. The hair examiner needs to know how a hair or hairs were collected and their precise location, along with suitable images to show location where this could be important. Where a hair examiner is given a hair or hairs without any information this would limit the interpretation and conclusions that a hair examiner may be able to make.

Hair examiners also need to work closely with relevant medical professionals to ensure that the collection of hairs is properly considered. This includes professionals at emergency areas in hospitals, specialist forensic medical centres such as sexual assault referral centres and morgues and forensic pathologists. Often the success or failure of a hair examination will depend on suitable known samples of hair having been taken during a medical procedure.

6.4.2 Analysis and Comparison of Material

The report should contain all relevant information about the receipt of items and a statement about subsequent storage and movement within the laboratory. The report may refer to these detailed records being available but not included in the report.

The report should contain information relating to the packaging and condition of the item on receipt and appropriate detail describing the actual item including each item having a unique identifier.

The methods used in examination of hairs should be included in the report or in an appendix. The appendix may cover the general approach followed in the laboratory with further case-specific information included in the report. An example of such an appendix can be found as Appendix 6.1, Protocol for Hair Examination, in this chapter.

The reader of the report should have a clear picture of the work undertaken, what questions the examiner is attempting to answer and sufficient descriptions and detail to form the basis of any conclusions.

Known limitations of methods or procedures that may affect results must be clearly stated in reports or in attached appendices.

6.4.3 Reporting Conclusions and Opinions

Under AS 5388.4, a requirement of the laboratory quality system is to have documented policies on the reporting of opinions based on qualitative and quantitative data and results.

As we have discussed previously, the examination of hairs relies on observation and qualitative assessments, excluding of course the reporting of the results of DNA analysis. Hence, professional judgement is required in forming an opinion.

It should be clear to the recipient of a report the basis for that opinion, including details of the test or analysis conducted. The wording used in reports should not understate or overstate the value of the test or analysis and should also not underestimate or overestimate their certainty.

Any reasonable alternative explanations or opinions should be included together with reasons for their rejection or a lower ranking. The hair examiner needs to carefully consider what are the relevant questions they are attempting to answer. This may mean requesting additional known samples. It is important that case notes record the reasoning behind the examinations conducted or why an examination may have been deemed unnecessary.

There will be circumstances where it is not possible to exclude or refute a hypothesis. In these circumstances, it may be appropriate to offer an *inconclusive conclusion* that neither supports nor refutes a hypothesis or its alternative. AS 5388.4 states that in these circumstances the use of the words "uninformative" or "neither supports nor refutes" should be considered. With respect to hair examinations the report should explain the particular reasons that may result in an inconclusive conclusion. This could include limitations with the recovered hair(s), for example, only a fragment of hair or lacking definitive microscopic features, or it may be that the known hair samples for comparison were deficient. It should be recognised that an inconclusive finding is not an error if the conclusion is based on sound scientific reasoning. However, an inconclusive finding should not be used as an excuse for not reaching a conclusion when there is a reasonable scientific basis for so doing.

Where the conclusion and opinion reached is exclusionary this should be clearly stated. The scientific basis for an exclusion should be clear to the recipient of the report.

Where the examiner reaches a non-exclusionary conclusion the wording of such a conclusion requires careful consideration.

The outcomes of some hair examinations may provide information that does not directly go to the identity of the source of the recovered hair. The identity or source may be known or not be in dispute. For example, the relevant question may be the type of hair and damage to the hair where an implement has allegedly been used as a weapon or it may be a question based on whether there is evidence of a particular type of implement having been used in an incident. Another example might be to assess whether hair has been removed with a degree of force. In brief, the examination of recovered hairs may assist in answering a range of questions that may be relevant to an investigation that do not centre around the identity of the source of the hair.

Where the identity of the source of a recovered hair is the central question, it is incumbent on the examiner to consider how best to word a conclusion that balances the need to not overstate what can be achieved through microscopic examination but at the same time, where possible, gives some sense of its potential "weight and substance" if used as evidence. We will now turn to some detailed consideration of appropriate wording to address this aspect.

6.4.4 Wording Used in Reporting Hair Conclusions

As we have previously discussed, for many hair examinations, it is not necessary to determine the identity or to individualise where the primary purpose is to provide useful information either for intelligence purposes or to assist more directly in an investigation. In the so-called *hierarchy of propositions* (Cook *et al*, 1998), hair examinations of this type would equate to hair findings contributing at the *activity level*. *Source level* information deals with from where (or whom) the evidence came. It is this level to which we now turn our attention and where the reporting of hair findings is potentially the most contentious and challenging.

Crocker stated that "the greatest challenge faced by forensic hair examiners is to be able to leave the witness box with a feeling of assurance that members of a jury, or a judge acting alone, have the same appreciation as the examiner does of the proper level of significance to be given to hair evidence" (Crocker, 1991).

His 1991 paper reported the results of a study that looked at forensic statements and wording and how these were perceived by jurors and law professionals. The participants were presented with a number of possible positive and negative statements and were asked to rate these in terms of degrees of certainty. The degrees of certainty were as follows:

1. Positive (certain)—A is B
2. Positive (probable)—A is probably B
3. Possible—A could be B, or inconclusive
4. Negative (probable)—A is probably not B
5. Negative (certain)—A is not B

The positive test statements included "is consistent with". This term was considered positive (certain) by 42% of jurors and positive (probable) by 35% of jurors. Almost 18% of jurors thought that it meant possible. Professionals indicating 38% positive (certain) and 62% positive (probable). The term "could have come from" was interpreted by all professionals as meaning possible, and only one juror thought it meant positive (certain) and 6% positive (probable).

This study was a surprise to us as we had thought that *consistent with* was a very weak finding and certainly weaker than *could have come from*. This demonstrates that scientists are not always the best placed to evaluate how what they say will be perceived by jurors or lawyers. This is supported in the literature where Howes *et al* (2013) found that scientist conclusions were written at a level where some university education would be required for them to be readily understood and that, even with this overall level of education, they would be difficult to read if that education was not in science. As scientists we generally write for scientists or professionals and not the general public who make up juries.

AS 5388.4 offers some guidance in what are acceptable and what are not acceptable phrases and words to use in reports. Terms such as *may, possible* and *could not be excluded* should not be used where results in balance favour one explanation over another. They suggest that it is better to use terms such as *likely not* or *probable* for such findings. The use of the term *consistent with* is discouraged without some form of qualifying statement suggesting the weighting to be given to an opinion so expressed. In line with the findings of the Crocker (1991) study, it cannot be assumed that jurors or legal professionals will assign the same weight as that intended by the scientist.

In our view, the terms *consistent with*, *similar* and *match* should not be used in reporting hair findings. This leaves us in the difficult position of how to phrase hair findings and assign some understanding of significance. Consistent with our approach to hair examination of focusing on meaningful differences, we are happy with the phrases *could not have come from* or *can be excluded* where the finding is that a hair(s) could not have come from a nominated known source.

Where it is not possible to exclude, due to any one of a range of limitations with samples, we are also happy with the use of the phrase *no meaningful conclusion can be reached* with regard to inclusion or exclusion. However, it should be clear to the reader the basis for this conclusion.

This leaves us with how best to express an opinion where we have not excluded based on microscopic examination and comparison. Although not entirely happy with the phrase, we still recommend the use of a phrase such as *cannot exclude* or *could have come from*. In using this terminology, it must be made entirely clear that it is not possible to absolutely individualise a hair on the basis of microscopic examination nor is this claimed. Further, whilst we believe that experience informing professional judgement cannot be denied, there should be no sense in a report of claiming that a hair is somehow rare because an examiner claims not to have seen a "similar" hair previously.

Notwithstanding, we reject the notion, sometimes put to us when giving testimony, that all hairs of a similar colour are indistinguishable. As we have previously stated a hair should be considered as a package of potential information. Clearly, a visually white hair when examined microscopically will have much less information than a visually coloured hair, but this does not mean that there is no information in a white hair or there are no differences between some white hairs. Some so-called white hairs will still have small amounts of pigment present and they typically may display other microscopic features such a cortical fusi and cortical texture. At the other end of the scale a visually black hair may have so much pigment that it makes it impossible to see any other detail and, hence, such hairs are of low information content. For these reasons, much of the focus on microscopic examination has historically been on visually coloured Caucasian hairs. If we take for example a mid-brown coloured scalp hair of Caucasian origin, we are likely to see a range of microscopically observable features that provide the basis for hair comparisons. It would be misleading to suggest that all brown hairs are identical because that is simply not the case. The difficulty is that we are not in a position to place any numerical estimate on how rare a combination of features might be in the overall population. It is simply not possible to estimate the frequency of detailed microscopic features.

AS 5388.4 states that if information on the rarity or otherwise of any correspondences is available from a relevant database, such information and any limitations of the information should be included in the report. It goes on to say that if an assessment on rarity depends on the experience of the examiner, a verbal descriptor such as *common*, *rare* or *extremely rare* should be used. Importantly, it adds that the basis of the opinion should be outlined in the report as a qualifying statement.

We would caution against the use of terms such as *rare* or *extremely rare* for hair conclusions as this may simply mean that the examiner has not personally seen a feature on a previous occasion. Where examiners are not dealing with a large number of cases, the latter is increasingly likely!

6.4.5 The Role of Statistics and Verbal Scales for Hair Opinions

AS 5388.4 states that where statistical calculations or weightings are used to support opinions, the basis of the statistical calculations (including assumptions) should be fully explained in the report or supporting documentation. Normally, the application of a statistical approach would require data capable of forming a meaningful database.

In some parts of the world and jurisdictions, the use of the so-called *evaluative reporting* has been proposed as a way of providing the strength of findings for competing propositions (hypothesis). Although not strictly synonymous, evaluative reporting may also be referred to as *the likelihood ratio approach, logical thinking* or *Bayesian inference.*

Guidelines and position statements on evaluative reporting have been issued by National Institute of Forensic Science, Australia and New Zealand (Anon, 2017), the European Network of Forensic Science Institutes (Anon, 2015b) and the Royal Statistical Society (RSS) (Jackson *et al*, 2014).

National Institute of Forensic Sciences (NIFS) defines evaluative reporting as a formalised thought process that enables evaluation of scientific findings given two opposing (or competing) propositions. It is argued that if properly used, it has the additional benefit of minimising *cognitive bias* and that opinions can be updated in a logical way on receipt of new information. The evaluative process is visually described in Figure 6.1 of the NIFS report (Anon, 2017)—see Figure 6.1. Cognitive bias is further discussed in Section 6.4.6.

The evaluation stage in Figure 6.1 **may** (our emphasis) involve the assignment of a numerical value for the probability of the findings or may be non-numerical.

The NIFS guide lists the body of knowledge and experience as relevant to informing the evaluation process. The body of knowledge can include published literature, databases of characteristic frequencies, and knowledge of analysis limitations, and experience can include, professional experience or personal knowledge, beliefs and assumptions.

We would add a word of caution with regard to reliance on personal knowledge and beliefs unless these have been formally written and subject to testing or review. This does not necessarily require formal

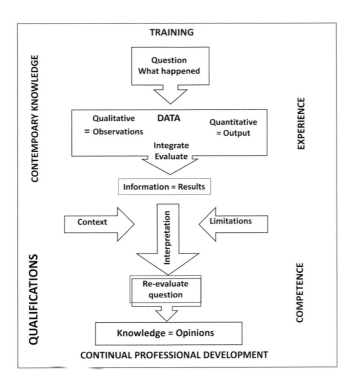

FIGURE 6.1 Role of professional judgement in evaluation and interpretation.

academic publication and may be in the form of formalised in-house validation studies.

We question the value of personal opinion where there is no demonstrated foundation for that opinion.

In the absence of a relevant population database or being unable to estimate feature frequencies in a reliable manner, as in the case of hairs, NIFS suggests that practitioners can utilise their experience and knowledge of the relevant population to *subjectively* estimate likelihood.

However, it should be clear and transparent in the report the basis for any conclusion and opinion reached based on the application of an evaluative process whether or not an actual likelihood ratio has been calculated. Indeed, the NIFS guide to evaluative reporting goes on to say that *practitioners may **avoid** (our emphasis) formal expression of a numeric probability and subjectively select a verbal qualifier to express a degree of support provide for the findings.*

The NIFS guide (Anon, 2017) gives the following example of a numerical and verbal scale.

Verbal conclusion	Likelihood ratio (LR)
Extremely strong support against	<0.000001
Very strong support against	0.000001–0.001
Strong support against	0.001–0.01
Moderate support against	0.01–0.1
Slight support against	0.1–1
Neutral	1
Slight support for	1–10
Moderate support for	1–100
Strong support for	100–1,000
Very strong support for	1,000–1,000,000
Extremely strong support for	>1,000,000

The use of verbal scales is not, however, without its own challenges and limitations.

Mullen *et al* (2014), in a survey of 400 volunteers, concluded that the terms used in verbal scales are unlikely to be understood properly by lay people with participants attributing a greater weight of evidence to lower weights of the verbal scale and lesser weight to the higher weights on the verbal scale. Martire *et al* (2014) also found a weak evidence effect and concluded that, given the high likelihood of miscommunication associated with low-strength verbal expressions of uncertainty, in instances where numerical values for low-strength evidence cannot be provided, it would seem appropriate to question whether expert opinions in the form of a verbal scale should be offered at all.

Martire and Watkins (2015) concluded that "if the intention of verbal conclusion scales is to facilitate effective and accurate communication of opinions regarding evidential weight, then that aim has not been achieved" and stated that "the high potential for miscommunication given the verbal conclusion scale is sobering".

Marquis *et al* (2016), whilst presenting arguments in favour of the use of verbal qualifiers, emphasise that they do not replace the use of numbers. However, they recognise that verbal qualifiers are prone to misunderstandings and cannot be coherently combined with other evidence. Hence, they recommend that verbal qualifiers are not used alone in reports and suggest that scientists should use examples to let the recipient of information understand how the scientific evidence affects the probabilities of propositions whilst also pointing out what the results do not mean. They express the opinion that if experts were able to coherently articulate numbers, and recipients could properly handle such numbers, then verbal qualifiers could be abandoned completely. We would suggest that this "ideal" is a long way off.

The NIFS guide (Anon, 2017) recognises that for some disciplines it may not be possible to consistently and accurately separate between levels on the scale and that a reduced scale can assist with consistency between practitioners. It also recognises that where the assignment of probabilities is based on practitioner judgement and experience, it is critical that within and between practitioner consistency is evaluated and checked. Within and between practitioner consistency for hair examiners has been the subject of concern and debate for many years (Gaudette, 1999; Robertson, 1982).

The potential to apply a Bayesian inference model to the evaluation and reporting of hair findings is also not a novel concept. Indeed, Gaudette proposed such an approach as long ago as 1999 and showed how his *average value of hairs* combined with factors that strengthen or weaken evidence could be incorporated in to a five-scale reporting scale. His reduced scale was as follows:

Strong positive: The questioned hairs originated from the same person as the known sample.

Normal positive: The questioned hairs are consistent with having originated from the same person as the known sample.

Inconclusive: No conclusion can be given as to whether the questioned and known hairs have a common origin.

Normal negative: The questioned hairs are not consistent with having originated from the same person as the known sample.

Strong negative: The questioned hairs could not have originated from the same person as the known sample.

His view was that the great majority of hair examinations would result in normal positive and normal negative conclusions. He stressed that the examiner should mention that hair comparison is not usually a positive means of identification. He proposed that the examiner would further evaluate the hair findings by considering a range of factors that tend to weaken or strengthen positive and negative conclusions. These are reproduced below. Note that we would now not recommend some of the terms used such as "similar" or "dissimilar".

Factors which tend to weaken positive hair comparison conclusions

1. The presence of incomplete hairs
2. Questioned hairs that are common featureless hairs
3. Hairs of non-Caucasian racial origin
4. A questioned hair found in conjunction with other unassociated hairs
5. Known samples with large intra-sample variation

Factors which tend to strengthen positive hair comparison conclusions

1. Two or more mutually dissimilar hairs found to be similar to a known sample
2. Hairs with unusual characteristics
3. Hairs found in unexpected places
4. Two-way transfer—for example, a victim's hair found on an accused's clothing and an accused's hair found on the victim's clothing
5. Additional examinations such as DNA analysis

Factors which tend to weaken normal negative hair comparison conclusions

1. Deficiencies in the known sample including, not enough hairs, not representative, contains incomplete hairs and large time difference between offence and procurement of known sample
2. Incomplete questioned hairs
3. Questioned hair has macroscopic characteristics close to those of the known sample

Factors which tend to strengthen normal negative hair comparison conclusions

1. Known sample has more than the recommended number of hairs
2. Known sample shows little intra-sample variation
3. Questioned hair has macroscopic and microscopic characteristics very dissimilar to those of the known sample
4. Two or more questioned hairs found together in a clump are dissimilar to the known sample

We would suggest that experts should not, as the NIFS guide suggest, default to a verbal scale to *avoid* the use of a numerical scale. In the absence of meaningful databases for hair examinations, it is doubtful that a Bayesian approach can currently sensibly be applied for hair findings.

6.4.6 Dealing with Cognitive Bias

Dror (2020) has discussed eight factors that can lead to bias and has proposed six ways in which bias can be minimised. Accepting that bias exists can be challenging as bias is sometimes misinterpreted as an imputation that in some way the scientist is behaving in an unethical way or may even be corrupt. Whilst there sadly have been instances of corrupt forensic scientists this is the exception. Dror points out that cognitive bias does not usually arise from dishonesty, intentional discrimination or a deliberate act arising from unethical behaviour but stresses that experts are not immune to bias and may not even be aware of their own biases. Indeed, experts may fall into the trap of believing that they can overcome bias by "mere willpower". This can in fact have the opposite effect and actually increase its effect through the so-called "ironic rebound" (Dror, 2020). Dror proposes the following six ways in which we can combat the various sources of bias.

1. Using blinding and masking techniques that prevent exposure to task irrelevant information.
2. Using methods such as linear sequential unmasking (LSU), to control the sequence, timing, and linearity of exposure to information, so as to minimise "going backward" and being biased by the reference materials.
3. Using case managers that screen and control what information is given to whom and when.
4. Using blind, double blind and proper verifications where possible.
5. Rather than having one hypothesis, having a "line up" of competing and alternative conclusions and hypotheses.
6. Adopting a differential diagnosis approach, where all different conclusions and their probability are presented, rather than one conclusion.

The challenge of mitigating for bias is a live issue for forensic organisations and each will over time, no doubt, develop their own approach and guidelines for practice and reporting. With respect to the examination of hairs, examiners should, at the very least, carefully consider alternative hypothesis and conclusions. This includes framing what are the relevant questions, determining appropriate examinations that are most likely to give insights to answer these questions, and considering reasonable alternative explanations that may assist the investigator, legal players and ultimately the fact finder be that an examining magistrate or juror.

6.4.7 Testimony and Giving Evidence

AS 5388-4 classifies testimony as a form of oral report.

The responsibilities of the scientist as an expert witness are often captured in professional or legal codes of practice for expert witnesses. These will differ in detail depending on the jurisdiction. A common and central element included in such codes is the overriding duty of the expert to assist the court impartially on matters relevant to their area of expertise and not as an advocate for any party.

As an expert you should be familiar with the specific requirements of your jurisdiction.

A normal witness is only allowed to speak to factual information and is not allowed to offer an opinion. Experts are allowed to offer opinions, indeed that is the reason why they are called as an expert witness. Most codes of practice or rules of evidence will spell out that such evidence must be based wholly or partly on a body of specialist knowledge. The expert is expected to be able to demonstrate that they have training, qualifications and experience relevant to that body of specialist knowledge. In leading evidence from an expert, the party calling the expert may ask questions to establish the latter, although in practice experts are not often asked to prove they have specialist knowledge unless the area of expertise is novel. It should go without saying, but it is worth a reminder, that an expert should not offer an opinion that is clearly outside their area of expertise and knowledge. On occasion an expert may be invited to stray beyond their area of expertise to assist the court—this is invariably not helpful!

In the continental or inquisitorial system an expert is usually appointed by the court and not by the parties (prosecution and defence). In the adversarial system, experts may be instructed by either the prosecution or the defence. None of this reduces the absolute requirement that the duty of an expert is to assist the court—there is no ownership of witnesses.

Where more than one expert is involved, they are expected to work cooperatively. Increasingly there are formal rules requiring that, if so directed, experts meet pretrial and discuss their findings and attempt to reach a joint position or at least be clear as to areas where there is disagreement.

It is beyond the scope of this book to consider all aspects of being an expert witness, the following is offered as a practical guide for the scientist facing their first or early appearances as an expert witness.

If you are employed by an organisation, their training program should include formal training on being an expert witness. This may include giving "evidence" in a moot or mock court simulation.

For academics or sole practitioners there are commercial training programs available in some countries. The court is a very unfamiliar place for the uninitiated!

Prospective expert witnesses should be familiar with the legal system under which they are appearing, including court protocol and procedure. It is always helpful to familiarise yourself with the physical layout of the court room including where the lawyers and jurors sit and the location of the witness box. All of this, of course, is just common sense but, nonetheless, needs to be planned and completed.

A training program should include the opportunity to attend court and observe more experienced experts giving evidence. Finally, it is important that you are dressed in a professional and presentable fashion so as to be seen as meeting appropriate professional standards.

The jurisdiction and legal system rule will determine how your evidence is taken. In the inquisitorial system, experts appear less frequently in person to present their findings. Indeed, even in jurisdictions where the adversarial system is in use experts appear on a less frequent basis than was previously the case. However, we will focus in the main on the process in an adversarial trial. It is now a rare event for an expert to be asked to give evidence in person at lower court hearings such as committal proceedings.

Firstly, you will be asked to either take an oath or to affirm that the evidence you will give will be the truth, the whole truth and nothing but the truth, or similar words. You should be clear whether you will take an oath or affirm.

The party calling you as a witness will lead your evidence in chief. Initially you can expect that you will be asked your name, for whom you work and some information about your qualifications—surprisingly our experience is that it is rare you will receive any detailed questioning about your qualifications and experience.

At this point your report may be shown to you and you will be asked to confirm that this is indeed your report. In some instances, the party calling you may tender your report and ask you no further questions. The "other" party may then accept your report or may cross examine you.

Where you are asked a question in chief you should pay close attention to the question you are asked and take time to think about your answer before speaking. Your answers should be concise and as unambiguous as possible. You should resist offering unnecessary information. You should speak slightly more slowly than normal conversation and articulate your words clearly.

In an ideal situation, the person asking you the questions will lead you through your evidence in a series of closed questions. You should resist trying to guess the next question in formulating your answer. However, the situation is not always ideal and there will be occasions where you will need to decide how much background information you need to give in order for the fact finder (jurors) to understand your evidence. Whilst your role is not to deliver an academic lecture you may have to lay the ground for the case specific information.

If you need to refer to your original notes or written report, you will need to ask permission to do so. In almost all circumstances you should seek to obtain such permission as early as possible when giving evidence and not rely on memory. It would be a very rare occasion that you were not given permission to look at your case file.

Whilst being polite and professional to the lawyers it is the jury, or a judge in a judge-only trial, to whom you should direct your answers. Usually, the witness box will be facing the jury allowing you to make eye contact with jurors which may assist you in assessing whether they are following what you are saying.

If you do not understand the question you are being asked, then say so—often a judge will support you, so don't be scared to politely ask for the question to be repeated or rephrased.

If you do not know the answer to a question, then it is better to say so, explain why if necessary, rather than attempt to answer the question. If you give an answer that you realise is wrong or potentially misleading, correct this answer as quickly as possible.

You may be surprised that you have not been asked about the information in your report. You should not volunteer information that you have not be asked for. There may be a sound reason why you have not been asked a question. You should not attempt to correct what you think is an error of "omission". If, however, you feel that a question may lead you to providing an answer that could be misleading, then it is your duty to attempt to correct an error of "commission". This may simply be that the lawyer asking the question does not have a sound grasp of the subject.

During giving evidence you may find yourself in a *voir dire* situation. This is where the opposing party challenges the question being asked. Often referred to as a "trial within a trial" the jury will normally be excluded whilst the two parties argue the point before the judge. In expert evidence this usually revolves around one party wishing to stop the other party asking a particular question or attempting to have evidence excluded. Once decided by the judge, the jury will be recalled and you will then be asked or not asked the question.

After your evidence in chief has been led, you may be cross examined by the opposing party and then re-examined (although this is rare). There are strict rules about what you can be asked in cross examination, but that is a matter for the lawyers.

On some occasions, and regrettably, parties will try and portray you as not being impartial. This may include suggesting directly, or by implication, that you owe some greater loyalty to your employer, especially if you work for a police organisation. If given the opportunity, you should emphasise that what is important is impartiality not independence from the party who has called you as an expert. You may wish to point out that everyone receives remuneration including the lawyers! If you work for an accredited organisation or hold some form of personal accreditation or formal professional status that includes adhering to a code of practice, you should point out the checks and balances that ensure your approach as a witness is impartial.

Your personal demeanour as a witness is critical in ensuring that you are seen as being an impartial witness. This includes careful use of language and not giving a sense of being defensive. The latter does not mean, however, that you should just accept whatever is put to you as this is not helpful to the fact finder. You should continue to support your conclusions and opinion whilst carefully considering alternative explanations for your findings.

Finally, you may leave the witness box not feeling especially comfortable with how your report/information was led, or after cross examination. This does not necessarily mean that you have not performed in an adequate manner as a witness. The adversarial system is called adversarial for a reason! However, you should be receptive to feedback on your performance as a witness.

Indeed, AS 5388-4 requires that organisations have a documented system of testimony monitoring. This normally requires some form of evaluation of the actual testimony. The later may include a structured feedback form competed by the parties (lawyers) and/or peer evaluation by a colleague. Evaluation of court transcripts may form part of an evaluation process but alone does not meet the requirement. In evaluating witness performance, it is really critical that this is seen as a part of continuous improvement to help you in your role as an expert witness.

APPENDIX 6.1: PROTOCOL FOR HAIR EXAMINATION

- The **known** (from a known source) sample(s) is (are) examined and five (or more) hairs are selected to represent the range of hair lengths and colours present. These hairs are placed on microscope slides in semi-permanent mountant, normally one complete hair to each slide. Features such as the profile or shape of the hair, length, colour and condition of the root and tip are then observed using a stereomicroscope and recorded. Using a comparison compound light microscope, detailed microscopic features may be recorded. An important part of this process is to study the variation in features along the length of each hair shaft.
- It is not always possible to adequately describe the features or variation present in a single hair using discontinuous classifications. A record sheet or features list is used to ensure systematic and thorough examination, but written descriptions are also used where appropriate.

- After completing an examination of the known hair sample(s), each recovered or questioned hair is examined separately in the same way as the known hair samples(s).
- Unless hairs have been excluded at the stereomicroscopic stage, each questioned hair is compared with one or more **known** hairs. These are selected on the basis of possessing similar features to the questioned hair. The comparison process involves looking for differences as well as features in common and comparing the pattern of features along the length of the hair shafts being compared.
- It is unlikely that a **questioned** and **known** hair will be indistinguishable for all features **along their entire length.** In order to conclude two hairs could have had a common origin, these hairs should show the same degree of variation and be indistinguishable at several points along their respective lengths. Any differences should be explicable in the forensic context.
- Microscopic examination can exclude hairs as having come from an individual. A conclusion that a hair or hairs could have come from an individual, an inclusion, does not usually mean that this would be the only person from whom the hair(s) could have originated. The strength of inclusion cannot be given a statistical estimate and can only be evaluated on a case-by-case basis.

APPENDIX 6.2: EXAMPLE REPORT 1

Name of Jurisdiction/Police Force

THE EXAMINATION OF A HAIR RELATING TO A COURT BOMBING, 2001–2002

BY

ELIZABETH MARY BROOKS

Senior Forensic Scientist

CONTENTS	PAGE
(1) CUSTODY OF ITEMS	1
(2) EXAMINATION AND RESULTS	1
(3) CONCLUSIONS	2
(4) HAIR APPENDIX	3
(5) REFERENCES	4

REFERENCE NUMBER:

1 March 2014

Page 1 of 4

This document is issued in accordance with NATA's accreditation requirements.

Accredited for compliance with ISO/IEC 17025 (Accreditation Number: .....).

1. Custody of Items

1.1 The following items were received by me Name of Jurisdiction/Police Force on the dates indicated:

Date Received	Item Number and Description
8 May2001	Case Reference Hair from blue material, FS11\23\02

2. Examination and Results

2.1 The purpose of my examination was to determine the source of the hair (Item 1) and if applicable its somatic origin.

2.2 I describe the approach and methodology used for forensic hair examinations in the Hair Appendix, Section 4.

2.3 I have examined and reviewed the single hair submitted to me for examination at low-power/stereo microscopic level and higher magnification with transmitted light microscopy, and I submit my general observations in Table 2.1 above.

2.4 High-power microscopy allowed cortical features of the hair to be observed. These included the uniform distribution of pigment, streaked aggregates of pigment apparent towards the tip end of the hair. The pigment granules themselves were fine, and whilst there were aggregates of melanin these did not present as ovoid bodies. The medulla was continuous, both opaque and translucent and cortical fusi were present and occasionally large along the length of the hair shaft.
Cortical texture was both obvious and coarse in places, while the cuticle was clear but damaged, particularly near the tip end. The hair shaft itself was damaged approximately 2.0 cm from the root end.

TABLE 2.1   Item 1—1× Hair from Blue Material, FS11\23\02

Item Number	Number of Hairs Examined	General Description	Examination Results
Item 1 – Hair from blue material, FS11\23\02	1	Microscopy examination of the single hair indicated that the hair was 5.0 cm long, was greyish brown at the root end that graded to a chocolate brown along the rest of the hair shaft. The hair had a telogen root type; some buckling and a prominent medulla were observed. The tip end of the hair was a natural taper.	Source: This is a human hair

3. Conclusions

3.1 The information described above allows the following observations to be made relating to Item 1—a single hair recovered from blue material:
- The hair is a human hair most likely comes from a body area such as the arm, chest, leg or back.
- It is a short hair (5.0 cm long), brown in colour with a telogen root type. A telogen root type on a human hair is important as it means this hair is not suitable for nuclear DNA analysis but could be suitable for mitochondrial DNA screening. Furthermore, a telogen root indicates that this hair was most likely lost naturally as it was in the resting phase of its growth cycle and is easily lost during the course of movement and activity—(most of the hairs collected in the course of crime scene investigations are hairs with telogen root types).
- The hair tip of Item 1 ends in a natural taper that means this hair has never been cut (as in barbering) and is probably at its final growth length.
- The cuticle and the hair shaft are damaged, but it is not possible to determine the causes of this damage.

3.2 It is my opinion that this hair was most likely to have been lost naturally in the course of daily activity and if a known sample of hair from a person can be obtained, a meaningful comparison can be made between the questioned hair and that known sample. Further to my opinion is that mitochondrial DNA analysis of this hair would allow a level of analytical DNA comparisons with other recovered DNA if present.

4. Hair Appendix

Protocol for Hair Examination

4.1 The known (from a known source) sample(s) is (are) examined and five (or more) hairs are selected to represent the range of hair lengths and colours present. These hairs are placed on microscope slides in semi-permanent mountant, normally one complete hair to each slide. Features such as the profile or shape of the hair, length, colour and condition of the root and tip are then observed using a stereomicroscope and recorded. Using a comparison compound light microscope, detailed microscopic features may be recorded. An important part of this process is to study the variation in features along the length of each hair shaft.

4.2 It is not always possible to adequately describe the features or variation present in a single hair using discontinuous classifications. A record sheet or features list is used to ensure systematic and thorough examination, but written descriptions are also used where appropriate.

4.3 After completing an examination of the known hair sample(s), each recovered or questioned hair is examined separately in the same way as the known hair samples(s).

4.4 Unless hairs have been excluded at the stereomicroscopic *stage,* each questioned hair is compared with one or more known hairs. These are selected on the basis of possessing similar features to the questioned hair. The comparison process involves looking for differences as well as features in common and comparing the pattern of features along the length of the hair shafts being compared.

4.5 It is unlikely that a questioned and known hair will be indistinguishable for all features along their entire length. In order to conclude two hairs could have had a common origin, these hairs should show the same degree of variation and be indistinguishable at several points along their respective lengths. Any differences should be explicable in the forensic context.

4.6 Microscopic examination can exclude hairs as having come from an individual. A conclusion that a hair or hairs could have come from an individual, an inclusion, does not usually mean that this would be the only person from whom the hair(s) could have originated. The strength of inclusion cannot be given a statistical estimate and can only be evaluated on a case by case basis.

5. References

Robertson, J 1999, 'Forensic and microscopic examination of human hairs', in Robertson, J (ed), *Forensic Examination of Hairs*, London: Taylor and Francis, pp. 79–154.

Brooks, EM & Robertson, J 2012, 'Natural and unnatural hair loss as detected in the forensic context', in Preedy, V (ed), *Handbook of Hair in Health and Disease*, The Netherlands: Wageningen Academic, pp. 216–235.

Gaudette, BD 1999, 'Evidential value of hair examination', in Robertson, J (ed), *Forensic Examination of Hairs*, London: Taylor and Francis, pp. 243–260.

Signed: Signature witnessed by:

APPENDIX 6.3: EXAMPLE REPORT 2

Name of Jurisdiction/Police Force Logo

**THE EXAMINATION OF A HAIR RELATING
TO THE MURDER OF MW 1983**

BY

Elizabeth Mary BROOKS

Senior Forensic Scientist

CONTENTS	*PAGE*
(1) CUSTODY OF ITEMS	2
(2) EXAMINATION AND RESULTS	2
(3) CONCLUSIONS	4
(4) HAIR APPENDIX	3
(5) REFERENCES	4

REFERENCE NUMBER:

1 March 2014 Page 1 of 4

This document is issued in accordance with NATA's accreditation requirements.

Accredited for compliance with ISO/IEC 17025 (Accreditation Number: ....).

1. Custody of Items

1.1 The following items were received by me:

Date Received	Item Number and Description
26 September 2013	Case Reference Number : Item 1 FS11/5321/6 Hairs from vacuuming's (boot of vehicle MAA 265)
26 September 2013	Case Reference Number : Item 2 FS83/435-6 Hairs from hairbrush—MW

2. Examination and Results

2.1 The purpose of my examination was to determine if any of the recovered hairs from **Item 1** could have come from the hairbrush MW—**Item 2**.

2.2 I describe the approach and methodology used for forensic hair examinations in the Hair Appendix, Section 4.

2.3 I have examined and reviewed a number of hairs submitted to me for examination at low power/stereo microscopic level and higher magnification with transmitted light microscopy and finally at high magnification comparative microscopy.

Comparative microscopy is the process whereby unknown or questioned hairs are compared at high magnification with hairs from a known source. This process attempts to establish if there are any observable differences between two hairs when viewed together in the comparison microscope.

All the hairs, fourteen (14) in total from **Item 1** (vacuuming's from boot of vehicle MAB 254) were mounted for microscopic examination. A summary of my observations is presented in Table 2.2 below:

2.4 The hairs KNOWN to have come from the hairbrush belonging to MW (**Item 2**) were visually assessed for length and colour variations and fourteen (14) hairs were mounted for examination by microscopy.

The hairs from **Item 2**, the hairbrush ranged, from 1.7 cm to 12.5 cm in length, tended to have a buckled appearance with the majority of the hairs, twelve (12), examined showing indications of artificial colouring, where multiple episodes of colouring were observed (eight [8] hairs had been coloured at least two [2] times). Artificial colours ranged between yellow to orange brown or black; however, hair colour associated

TABLE 2.2 Summary of Hairs Examined from Item 1—Vacuuming's from Boot of Vehicle MAB 254

Item 1 Summary Includes Hairs Items 1–1 to 1–14		Number of Hairs[a]
Colour	Colourless to yellow	1[c]
	Orange	1[c]
	Grey brown/brown	1
	Black/charcoal	2
Root appearance	Root (anagen/telogen)	3[b]
	Absent—cut	1
	Broken	1
Tip appearance	Natural taper	1
	Cut	3
	Broken	1
Somatic origin	Scalp hair	5
	Body hair (includes facial hair)	2
	Animal hair	2
	Origin—Not hairs	5

[a] Items that were animal hairs or "not hairs" were not included in the descriptions of hair colour, root and tip appearance that here accounts for nine (9) of the total fourteen (14) items examined.
[b] Hairs with telogen growth phase roots.
[c] Hair has been artificially coloured.

with natural regrowth tended to be generally grey brown and occasionally a charcoal colour. Most hairs recovered for mounting from **Item 2**—the hairbrush, had telogen stage root growth, and had either cut hair tips or natural tapers.

3. Conclusions

3.1 Extensive comparative microscopy was undertaken between **Item 1–3, Item 1–9 and Item 1–14** recovered hairs from **Item 1,** with hairs KNOWN to have come from the hairbrush belonging to MW (**Item 2**) specifically, **Item 2–4 and Item 2–6.**

3.2 In the case of the comparison between the known hairs, **Items 2–4 and Item 2–6** and three (3) questioned hairs **Item 1–3, Item 1–9 and Item 1–14** no observable differences could be detected.

3.3 It is my opinion that the three questioned hairs from **Item 1 (Item 1–3, Item 1–9 and Item 1–14)** could have come from the same person whose hairs were retrieved from the hairbrush, **Item 2,** and known to have come from MW.

4. Hair Appendix

Protocol for Hair Examination

4.1 The **known** (from a known source) sample(s) is (are) examined and five (or more) hairs are selected to represent the range of hair lengths and colours present. These hairs are placed on microscope slides in semi-permanent mountant, normally one complete hair to each slide. Features such as the profile or shape of the hair, length, colour and condition of the root and tip are then observed using a stereomicroscope and recorded. Using a comparison compound light microscope, detailed microscopic features may be recorded. An important part of this process is to study the variation in features along the length of each hair shaft.

4.2 It is not always possible to adequately describe the features or variation present in a single hair using discontinuous classifications. A record sheet or features list is used to ensure systematic and thorough examination, but written descriptions are also used where appropriate.

4.3 After completing an examination of the known hair sample(s), each recovered or questioned hair is examined separately in the same way as the known hair samples(s).

4.4 Unless hairs have been excluded at the stereomicroscopic stage, each questioned hair is compared with one or more **known** hairs. These are selected on the basis of possessing similar features to the questioned hair. The comparison process involves looking for differences as well as features in common and comparing the pattern of features along the length of the hair shafts being compared.

4.5 It is unlikely that a **questioned** and **known** hair will be indistinguishable for all features **along their entire length.** In order to conclude two hairs could have had a common origin, these hairs should show the same degree of variation and be indistinguishable at several points along their respective lengths. Any differences should be explicable in the forensic context.

4.6 Microscopic examination can exclude hairs as having come from an individual. A conclusion that a hair or hairs could have come from an individual, an inclusion, does not usually mean that this would be the only person from whom the hair(s) could have originated. The strength of inclusion cannot be given a statistical estimate and can only be evaluated on a case by case basis.

5. References

Robertson, J 1999, 'Forensic and microscopic examination of human hairs', in Robertson, J (ed), *Forensic Examination of Hairs*, London: Taylor and Francis, pp. 79–154.

Brooks, EM & Robertson, J 2012, 'Natural and unnatural hair loss as detected in the forensic context', in Preedy, V (ed), *Handbook of Hair in Health and Disease*, The Netherlands: Wageningen Academic, pp. 216–235.

Hicks, JW 1977, Microscopy of hairs. A practical guide and Manual, Issue 2, Washington, DC: Federal Bureau of Investigation.

Signed: Signature witnessed by:

APPENDIX 6.4: EXAMPLE REPORT 3

Name of Jurisdiction/Police Force Logo

THE EXAMINATION OF A HAIR RELATING TO A MOTOR VEHICLE ACCIDENT INVOLVING TWO OCCUPANTS—X AND Y

BY

JAMES ROBERTSON BSc PhD

Forensic Scientist

CONTENTS	PAGE
(1) CUSTODY OF ITEMS	1
(2) EXAMINATION AND RESULTS	2
(3) CONCLUSIONS	2
(4) HAIR APPENDIX	3
(5) REFERENCES	4

REFERENCE NUMBER: ....

1 March 2014

This report consists of four (4) pages each signed by me

This document is issued in accordance with NATA's accreditation requirements.

Accredited for compliance with ISO/IEC 17025 (Accreditation Number: ....).

1. Custody of Items

1.1 I received the following items on 3 February 2014 from .... of the Police, Forensic Services:

Date Received	Item Number and Description
3 February 2014	Case Reference Number : Item 1 1× yellow top container with hair samples from windscreen
3 February 2014	Case Reference Number : Item 2 Scalp hair sample from ....Y
3 February 2014	Case Reference Number : Item 3 Scalp hair sample from X

2. Examination and Results

2.1 I received scalp hair samples from two individuals, X and Y, and two hairs which I understand to have been recovered from the windscreen of a vehicle involved in an accident.

2.2 I have examined the above samples following the approach and methodology outlined in the attached appendix.

2.3 The known SCALP HAIR sample from X consisted of over 50 hairs ranging in length from about 10 cm to 38 cm. Overall the hairs were a mid to dark brown in colour. Individual hairs varied in colour from grey to almost black near the root end becoming a truer brown along the hair shaft. In one hair there was a profound and sharp change in colour from very dark brown to yellow indicative of a dyed hair. Other hairs may have had artificial colour treatment. Ten hairs were prepared for detailed microscopic examination.

2.4 The known SCALP HAIR sample from Y consisted of over 50 hairs ranging from about 16 cm to 29 cm. Overall the hairs were a dark brown in colour. Examination at low magnification revealed that there was a profound sharp change in colour along individual hair shafts from grey to dark brown. The hairs from Y had been dyed. Ten hairs were prepared for microscopic examination.

2.5 Against this background, I examined two hairs which I understand were recovered from the windscreen of a vehicle involved in an accident. These hairs were in fact two segments of hair shaft, respectively, 7 cm and 6 cm in length. The longer hair had a cut end and a broken end and was a pale to light brown in colour, the shorter hair was broken at both ends and was dark brown in colour. Both hair fragments were prepared for detailed microscopic examination.

3. Conclusions

3.1 After detailed examination of the two known hair samples and the recovered hairs and direct comparisons, I reached the following conclusions.

The pale recovered hair is outside the range of colour of the two known samples and, hence, I can reach no conclusion as to its origin other than to state it is of human origin and scalp hair.

The darker recovered hair can be distinguished from the hairs of Y and, in my opinion did not originate from this person. The darker hair did not show any forensically significant differences from parts of hairs from X and, in my opinion, could have been the source of this hair fragment. The recovered hair was a fragment. It did not display the full range of features seen on the complete hairs of X. Under normal circumstances this would increase the chance that someone other than X could have been the source of this hair. However, if the only persons having access to the vehicle were Y and X, then the latter is not relevant. The recovered darker hair had broken ends and is not likely to have been shed during normal activities.

4. Hair Appendix

Protocol for Hair Examination

4.1 The **known** (from a known source) sample(s) is (are) examined and five (or more) hairs are selected to represent the range of hair lengths and colours present. These hairs are placed on microscope slides in semi-permanent mountant, normally one complete hair to each slide. Features such as the profile or shape of the hair, length, colour and condition of the root and tip are then observed using a stereomicroscope and recorded. Using a comparison compound light microscope, detailed microscopic features may be recorded. An important part of this process is to study the variation in features along the length of each hair shaft.

4.2 It is not always possible to adequately describe the features or variation present in a single hair using discontinuous classifications. A record sheet or features list is used to ensure systematic and thorough examination, but written descriptions are also used where appropriate.

4.3 After completing an examination of the known hair sample(s), each recovered or questioned hair is examined separately in the same way as the known hair samples(s).

4.4 Unless hairs have been excluded at the stereomicroscopic stage, each questioned hair is compared with one or more **known** hairs. These are selected on the basis of possessing similar features to the questioned hair. The comparison process involves looking for differences as well as features in common and comparing the pattern of features along the length of the hair shafts being compared.

4.5 It is unlikely that a **questioned** and **known** hair will be indistinguishable for all features **along their entire length.** In order to conclude, two hairs could have had a common origin, these hairs should show the same degree of variation and be indistinguishable at several points along their respective lengths. Any differences should be explicable in the forensic context.

4.6 Microscopic examination can exclude hairs as having come from an individual. A conclusion that a hair or hairs could have come from an individual, an inclusion, does not usually mean that this would be the only person from whom the hair(s) could have originated. The strength of inclusion cannot be given a statistical estimate and can only be evaluated on a case by case basis.

5. References

Robertson, J 1999, 'Forensic and microscopic examination of human hairs', in Robertson, J (ed), *Forensic Examination of Hairs*, London: Taylor and Francis, pp. 79–154.

Brooks, EM & Robertson, J 2012, 'Natural and unnatural hair loss as detected in the forensic context', in Preedy, V (ed), *Handbook of Hair in Health and Disease,* The Netherlands: Wageningen Academic, pp. 216–235.

Hicks, JW 1977, Microscopy of hairs. A practical guide and Manual, Issue 2, Washington, DC: Federal Bureau of Investigation.

Signed: Signature witnessed by:

CHAPTER 7

Training Considerations

7.1 SCOPE

A requirement for accreditation against ISO/IEC 17025:2018 includes that the laboratory shall have procedures and retain records for training, authorisation and monitoring competence of personnel. Each organisation must also have a management system that documents its activities including its protocols and methods.

The Scientific Working Group for Material Science Analysis (SWGMAT) has issued guidelines for forensic human hair identification and comparison (Anon, 2005). The Organisation of Scientific Area Committees for Forensic Science (OSAC), under the National Institute of Standards and Technology (NIST) in the United States of America, has sent these guidelines to ASTM International for development as an ASTM Standard. OSAC have also sent guidelines for the "Standard Practice for Training in the Forensic Examination of Hair by Microscopy" for development as an ASTM Standard. The European Network of Forensic Science Institutes (ENFSI) has also produced a manual of best practice for hair examination (Anon, 2015c). The OSAC and ENSFI documents would assist an organisation in developing its own training manual.

The ENSFI Fibre and Hair expert group have also developed a training and education e-platform (https://e-learning.ethg.eu) covering four main areas. These are as follows:

1. *From crime scene to the laboratory:* Case assessment, packaging and documentation, recovery methods, anti-contamination and sampling strategies.
2. *Microscopy:* Fundamental concepts of the important microscopic techniques as well as guidance tools for the discrimination, identification and comparison of fibres and hairs.
3. *Microspectrophotometry (MPS):* Fundamental concepts, sampling and analysis and interpretation of spectra.
4. *Fourier transform infrared spectroscopy (FTIR):* fundamental concepts and sample preparation.

The SWGMAT hair guidelines include the requirement for a formal training programme covering principles, methods and procedures applied to hair comparisons.

The level of detail in a hair training manual should capture the scope of testing conducted by that organisation. It serves no useful purpose for an organisation to document tests it has no intention to conduct. As we have discussed throughout this book, the scope of hair examinations can vary from very preliminary screening of hairs through to detailed microscopic examination and comparison. Hence, in this chapter we present guidance to develop a training programme aimed at examiners to conduct tests at the first three levels of examination that we have described throughout this book, that is, level 1, recognition and separation of human and animal hair; level 2, examination of human hairs including body area and ethnic origin and selection of hairs for DNA and level 3, detailed examination of hairs and comparison microscopy. Level 4 covers specialist areas of hair examination and is outside the scope of this book.

This approach means that an organisation need not waste time and energy in providing training that is outside of the scope of the tests they conduct. The skills and knowledge at each level build on each other in a sequential approach so that examiners can be authorised to conduct tests and report at each level as they demonstrate competence.

7.2 ASSUMED KNOWLEDGE AND COMPETENCIES

Our training guideline assumes that participants will have completed an organisation induction programme, will be familiar with their organisation's quality system and have started in a discipline-specific area. We have used AS 5388 parts 1–4 as a framework throughout this book. Hence, participants would be expected to have a good working knowledge of this standard or a similar standard adopted by their organisation. As AS 5388 is discipline agnostic, each discipline will have to interpret its requirements for their discipline.

Participants in a hair training programme may come from a number of disciplines including crime scene, biology and chemistry especially where such individuals have the responsibility for the examination of hairs in a triaged case management system post crime scene collection or in a laboratory setting.

Hence, it is also assumed that participants will have received *general* training in the recognition, recording and collection of physical materials and on evidence continuity and packaging. Hair specific considerations can then be layered on this general knowledge.

Microscopy plays a central role in the examination of hairs. Hence, participants are expected to have competencies in using low-power microscopic examination (LPM) for levels 1 and 2 and transmitted light microscopy (TLM) for level 3 examinations. Level 2 may require the use of TLM.

Following any level of examination, some form of case report would be expected. The interpretation and reporting of hair findings is again level dependent. However, we will not cover generic training in interpretation, reporting and on giving evidence which should be included in any forensic training programme. Some guidance is offered on interpretation and reporting of hair findings in Chapters 6 and 7 of this book.

At an organisational level, the organisation **must have** a relevant reference collection of hair samples. The scope of this collection should include the type of hairs that might reasonably be encountered in routine case work. Records should be kept that demonstrate that samples have a provenance such that they can be relied upon as representative of the species they are said to represent. Unfortunately, many organisations have collections built up over many years that do not meet this standard— these should not be used for training purposes.

7.3 LEVEL 1 TRAINING—RECOGNITION AND SEPARATION OF HUMAN AND ANIMAL HAIRS

7.3.1 Overview

The ability to recognise, effectively and efficiently recover hairs and to separate human from animal hairs are basic skills that should reside in any organisation conducting evidence recovery and basic hair examinations. This training guideline aims to provide participants with the knowledge and skills to conduct this level of examination and for an organisation to authorise participants to conduct such tasks. Level 1 training consists of four modules.

7.3.2 Module 1

Recovery and collection of hairs.

7.3.2.1 Learning Outcomes

On completion of this module participants will be able to:

- Understand and describe the factors that influence the transfer and persistence of hairs.
- Understand and describe the factors that determine how hairs should be recovered, collected and stored.

- Demonstrate a range of techniques and approaches to hair recovery and collection.
- Understand and describe the factors that determine what are adequate and representative known (exemplar) and reference samples.
- Understand and discuss contamination, preservation, continuity and security issues as they relate to the integrity of hair as evidence

7.3.2.2 Content

Content should include both theoretical and practical aspects. Chapter 3 and parts of Chapter 4 of this book provide essential foundational theory. SWGMAT trace Evidence Recovery Guidelines also provides relevant foundational knowledge.

Participants should research the relevant literature for specific scientific papers and be able to present and discuss one or more such papers in a group setting.

Practical exercises may involve recovery of hairs from mock crime scenes and/or "seeded" exhibits in a laboratory setting *following the authorised methods of the organisation*. However, it is also important to consider other methods not used by an organisation and understand the pros and cons of alternative approaches.

Participants should be shown how to take known samples of human and animal hairs. Where ethical approval is available, participants may collect known hairs from themselves and donors with informed consent.

The use of case studies, including wrongful convictions in which hair evidence has played a part, are important to understand best practice.

7.3.2.3 Competency Evaluation

Evaluation should include formal tests/quizzes on theoretical knowledge and a range of practical assessments which may include a capstone exercise where the trainee has to demonstrate competence in meeting all requirements of evidence recovery for hairs from a number of relevant mock case scenarios.

7.3.3 Module 2

Introduction to the biology and chemistry of hairs.

7.3.3.1 Learning Outcomes

On completion of this module participants will be able to:

- Understand at an introductory level, the biology and chemistry of hair.
- Understand and describe how hair grows.

7.3.3.2 Content

This theory-based module should provide knowledge of the biology and chemistry of hair at an introductory or foundational level. Chapter 2 is required reading and covers the structure of the hair follicle, the hair growth cycle, hair growth rates and hair distribution, and the basic structure or morphology and anatomy of hair. This module also lays the foundation knowledge for level 2 and level 3 training.

7.3.3.3 Competency Evaluation

Participants should demonstrate a working knowledge of basic hair biology and chemistry. This may be assessed through multiple choice/quiz type questions and participants making presentations on a specific aspect of hair biology and/or chemistry following their own literature research.

7.3.4 Module 3

Principles of microscopy and sample preparation.

7.3.4.1 Learning Outcomes

On completion of this module participants will be able to:

- Demonstrate an understanding of the theory of LPM and TLM.
- Understand the constituent parts and the appropriate theory underlying the use of microscopes.
- Understand and be able to discuss the factors to be considered in the use of microscopy in the hair examination protocol.
- Demonstrate a range of techniques in the preparation of hairs for microscopic examination.
- Understand the factors determining choice of suitable mountants for hair microscopy.

7.3.4.2 Content

The purpose of this module is to ensure that individuals responsible for the recovery of hairs understand at what point in the examination of recovered hair the use of microscopy should be considered and to understand the limits of what can be seen with visual examination alone. The module should also provide a basic introduction to LPM and TLM and the preparation of hairs for examination.

Theory should include an introduction to microscope construction, component parts, nomenclature and when it is appropriate to use LPM and TLM. Participants are expected to be microscope literate and understand the underlying theory of microscopes.

At this level, participants should understand the factors which need to be considered in deciding whether to mount a hair and, if so, factors influencing the choice of mountant.

Participants should demonstrate competence in mounting hairs and the use of LPM.

At this level, participants would not be expected to demonstrate advanced theoretical knowledge of TLM or practical skills in the use of TLM.

7.3.4.3 Competency Evaluation

Participants should be able to demonstrate a theoretical understanding of how and when to use LPM, when and how to mount hairs and the use of LPM for preliminary examinations of hairs. Participants should demonstrate practical skills in mounting hairs and in the use of LPM.

7.3.5 Module 4

Structure of human and animal hairs—differentiating human and animal hairs.

7.3.5.1 Learning Outcomes

On completion of this module participants will be able to:

- Understand and describe the morphological structure of a hair.
- Understand the types of microscopic features at LPM used to describe animal hairs.
- Understand the types of microscopic features at LPM used to describe human hairs.
- Understand the features used to identify a hair as human or of animal origin.
- Understand the protocols for differentiating human and animal hair including the use of checklists where appropriate.
- Demonstrate competence in separating co-mixed animal and human hair at the LPM level.

7.3.5.2 Content

Chapter 3 and parts of Chapter 4 of this book provide the theoretical underpinning of this module.

Participants should examine a range of human and animal hairs, separately and admixed to simulate typical case scenarios. Animal hairs selected for examination should reflect hairs typically encountered in case work. Participants should follow the hair examination protocol of their organisation. This should follow a logical sequence of visual examination followed by LPM of unmounted hairs or mounted hairs.

Checklists should be used to ensure a thorough and systematic recording of each hair.

7.3.5.3 Competency Evaluation

As in previous modules this should include a theory test which may comprise a mix of short questions and quizzes to test participant knowledge.

As this is the culminating practical competency for level 1 examinations there must be a comprehensive competency test that tests participants ability to effectively separate human from animal hairs. This may involve a number of simulated case scenarios. Participants will not be expected to be able to identify animal hairs.

7.4 LEVEL 2 TRAINING—EXAMINATION OF HUMAN HAIR INCLUDING BODY AREA AND ETHNIC ORIGIN AND SELECTION OF HAIRS FOR DNA TESTING

7.4.1 Overview

Once a hair or hairs have been recovered and identified as being of human origin, the next step in the examination process should be to determine the body area origin and ethnic origin of the recovered hair. This requires that recovered hairs are examined at LPM level and may require some preliminary examination at TLM level.

Selection of hairs suitable for routine nu-DNA testing requires the examination of the root end of hairs and, where a root is present, a determination of the growth phase of the root and its potential for DNA analysis.

Level 2 training assumes that participants have successfully completed level 1 training and have achieved competency at this level. Hence, level 2 training builds on level 1 training and adds new skills and competencies. Level 2 training consists of three modules.

7.4.2 Module 1

Body area determination.

7.4.2.1 Learning Outcomes

On completion of this module participants will be able to:

- Describe the features of hairs from different body areas.
- Demonstrate competency in classifying hairs on the basis of body area origin.
- Understand the limitations of body area determinations and implications for reporting.

7.4.2.2 Content

Chapter 4 of this book provides the theoretical underpinning of this module.

Participants should examine a range of hairs from different body areas including scalp, pubic, facial and other body hairs. Hairs from these locations should be sourced from individuals identifying as having different ethnic origin. At a minimum, hair types to be examined need to reflect those typically seen in routine case work. Body area determinations are made using LPM, with the use of a checklist with supplementary drawing, images and descriptive notes. The use of TLM may assist in the determination of body area but it does not generally require recording of detailed microscopic features at this level of microscopy.

7.4.2.3 Competency Evaluation

Participants should demonstrate their understanding of the features of value in determining body area through a theory test which may comprise short questions and/or quizzes. Questions should establish that participants understand the limitations in determining body area origin and how to report such findings.

Competency should be tested through participants having successfully classify a number (at least 20 hairs) of questioned hairs of known origin. In our view, it is not possible to accurately sub classify general body hairs into leg, arm or other body locations. Hence, the test should be set up at the level of scalp, pubic, facial (including eyebrow, eyelash and beard hair) and body hair from individuals identifying as having different ethnic origin.

7.4.3 Module 2

Ethnic origin determinations.

7.4.3.1 Learning Outcomes

On completion of this module participants will be able to:

- Describe the features of hairs from persons of different ethnic origin.
- Demonstrate competency in classifying hairs on the basis of ethnic origin.
- Understand the limitations of ethnic origin determination and implications for reporting.

7.4.3.2 Content

Chapter 4 of this book provides the theoretical underpinning of this module.

Initial examination may be limited to scalp hairs from individuals who identify as European, Asian and African. However, in order to appreciate the limitations of assigning ethnicity through hair examination participants should examine a range of hairs from different body areas and from individuals who identify as coming from mixed ethnic origins. Generally, ethnic determinations can be made at the LPM level of microscopic examination complemented where necessary with TLM examination. Features should be recorded using checklists with supplementary drawings, images and descriptive notes.

7.4.3.3 Competency Evaluation

The determination of ethnic origin is controversial, and some commentators question its validity and even its appropriateness in today's society. Hence, an important element for participants is to demonstrate that they understand the contemporary issues in this area of hair examination. This may be tested through formal questions and in-depth discussion of the topic.

In our view, participants should be tested with assigning hairs to an ethnic group only to demonstrate the difficulty in doing so except into the broadest classification.

7.4.4 Module 3

Selection of hairs for nu-DNA testing.
On completion of this module participants will be able to:

- Understand the current status of nuclear and mitochondrial DNA testing in relation to hairs.
- Understand the factors to be considered in selecting hairs for DNA testing in the context of the contemporary management of cases involving hairs.
- Understand and recognise the growth phases of hair through the appearance of hair roots.
- Understand the criteria for selecting hairs for non-routine nu-DNA testing.
- Understand the criteria for selecting hairs for mt-DNA testing.
- Demonstrate competency in the selection of hairs suitable for routine nu-DNA testing.
- Where in the scope of a laboratory demonstrate competency in the techniques used to select hairs for non-routine nu-DNA testing.

7.4.4.1 Content

Chapters 1, 2 and 4 are required reading for this module. The selection of hairs for DNA is dependent on the knowledge of hair cycle and the ability to correctly identify the growth phase of a hair based on the appearance off the hair root. Content, therefore, needs to include sufficient under-pinning knowledge of the biology of hair growth and the hair growth cycle. Each organisation should have guidelines or rules on the selection of hairs for routine nu-DNA testing and for mt-DNA testing. Such guide-lines should recognise the need to balance the potential for type 1 and type 2 errors. Some organisations will also include guidelines for non-routine nu-DNA testing. Participants may not be forensic biologists and a detailed knowledge of DNA testing is not required for this module. Participants should have the knowledge of the current status of nuclear and mitochondrial DNA as they apply to hairs as a biological source.

7.4.4.2 Competency Evaluation

Participants should be tested for their understanding of underlying the-ory through short questions and quizzes.

The key competency for this module is the ability to select hairs for routine nu-DNA testing based on root appearance. Depending on the organisation hairs may be examined unmounted or mounted. Participants should examine at least 100 hair roots representing the three growth stages, anagen, catagen and telogen and achieve at least 95% correct assignment as telogen and non-telogen. Only telogen hairs are not suitable for routine nu-DNA testing.

For those organisations that undertake non-routine nu-DNA test-ing participants should demonstrate competence in the application of a nuclear staining method and interpretation of the appearance of stained hair roots. As science progresses new approaches to the analysis of DNA in hair shafts may be developed and be introduced in to forensic procedures.

Note, module 3 may be delivered as a standalone module at level 1 where an organisation does not include body area and/or ethnic determinations.

7.5 LEVEL 3 TRAINING—DETAILED EXAMINATION OF HAIRS AND COMPARISON MICROSCOPY

7.5.1 Overview

For many organisations hairs will not be examined beyond level 1 and 2 and may even be limited to selection of hairs only for routine nu-DNA testing.

Level 3 examination should only be undertaken by organisations who see value in the more detailed microscopic examination of hairs as part of their triage and management of cases involving hairs.

At this level, hairs **must be mounted** in a suitable mountant and the laboratory must have access to high quality TLM. For comparison microscopy, the organisation must have a suitable comparison microscope. Participants at this level must have successfully completed level 1 and 2 training and demonstrated their competence.

Depending on how level 1 and 2 training is delivered, it is reasonable to estimate that such training could be completed in a matter of days or weeks with intensive practical sessions. By contrast level 3 training is a much larger commitment and could extend over a period of many months or even years. Hence, level 3 training should not be entered into lightly and requires long term commitment on the part of participants and the organisation.

Organisations need to be clear as to what they see as value deriving from this level of training which should never be seen as leading to identification. Even at this level, the primary outcome of hair examinations will be exclusions based on meaningful differences and to select the most suitable hairs for mt-DNA testing where available.

Prerequisites for participants are that they

- have completed level 1 and 2 training as this provides the underpinning knowledge on which level 3 training is built,
- have been authorised to use LPM and TLM,
- have read and are familiar with the content of all chapters of this book or equivalent theory,
- have access to a reference collection of hairs that includes examples of hairs likely to be encountered in their jurisdiction, and
- have access to high quality LPM, TLM and a comparison microscope.

The suggested structure for level 3 training is for five modules run over a minimum of 6–12 months although there is no fixed time scale and training may be extended over a longer time period. The time allocated to trainees must be sufficient for participants to conduct extensive practical laboratory work and to complete detailed proficiency testing.

7.5.2 Module 1

Developing advanced knowledge of the biology and chemistry of hair.
On completion of this module participants will be able to

- Demonstrate an understanding of the biology and chemistry of hairs appropriate to the underpinning knowledge required for the detailed examination of hairs.

7.5.2.1 Content

Participants will have completed level 1 and level 2 training and will have a foundation level knowledge. The content of this module should build on that foundational knowledge to ensure that participants have a deeper knowledge of the structure of the hair follicle, the growth cycle for hairs, hair distribution and growth rates and the anatomy and morphology of hairs. As this is a theory-based module participants should be familiar with Chapters 2 and 4 of this book and would be expected to conduct independent research on specific aspects of hair biology and/or chemistry.

7.5.2.2 Competency Evaluation

There are no practical competencies for this module. Participants should be tested on their knowledge through written and oral presentations, including the preparation of in-depth research-based papers and presentations to peer groups.

7.5.3 Module 2

Developing an advanced knowledge of the microscopic features of hair at the LPM level.

On completion of this module participants will be able to:

- Demonstrate an in-depth understanding of the features that can be assessed at the LPM level.
- Demonstrate an ability to classify and accurately record the features that can be assessed at the LPM level.
- Demonstrate the ability to correctly exclude recovered/questioned hairs from known samples at the LPM level.
- Understand both inherent and acquired characteristics of hair as seen at the LPM level.
- Understand the limitations of hair examination at the LPM level of examination.
- Write reports on outcomes of hair examinations at the LPM level.

7.5.3.1 Content

Participants should understand the criteria for the selection of features assessed by their organisation to examine hairs at the LPM level. Chapter 4 of this book and Chapter 2 of Robertson, 1999, *Forensic Examination of Hair*, Taylor & Francis Group, provide a suitable level of theory and underpinning knowledge for this module. Chapter 8 of this book discusses acquired characteristics.

In addition to underpinning theory participants should understand what can properly be concluded and what should not be concluded from LPM examinations. This includes an appreciation of type 1 and type 2 errors, the comparison process, the criminalistics value of features and the limitations of hair examination at the LPM level of examination.

7.5.3.2 Competency Evaluation

Participants should be tested on their theoretical knowledge through short questions and/or quizzes.

Participants should be tested on their practical skills through structured practical exercises. Feature recognition can be tested with the use of authenticated images of microscopic features but must also include the examination of a range of hairs selected to show the range of microscopic features seen at LPM.

Participants should classify and record features using a checklist and written descriptions as appropriate. Whilst there is no minimum or maximum number of hairs that should be examined, there should be sufficient samples to provide confidence that participants demonstrate genuine competency through demonstrating a very high degree of success in accurately classifying and recording the potential information available for each hair.

A useful practical exercise is for participants to build their own set of reference samples and slides by simulating damage to hairs using a range of cutting implements and other simulated damage such as crushing.

The capstone exercise for this module should be a series of mock case scenarios. Participants should demonstrate that they can reach the correct conclusion for each scenario and write a suitable report for each scenario demonstrating that they understand the limitations of hair examination at the LPM level.

7.5.4 Module 3

Developing an advanced knowledge of the microscopic features of hair at the TLM level.

On completion of this module participants will be able to:

- Demonstrate an in-depth understanding of the microscopic features that can be assessed at the TLM level.
- Demonstrate an ability to classify and accurately record the microscopic features that can be assessed at the TLM level.
- Demonstrate an understanding of variation in microscopic features within and between hairs at the TLM level.

- Demonstrate the ability to correctly exclude questioned/recovered hairs from known samples at the TLM level.
- Understand the limitations of hair examination at the TLM level.
- Write reports on the outcomes of hair examinations at the TLM level.

7.5.4.1 Content

A prerequisite skill for this module is that participants are competent in the use of TLM. Content may include revision exercises to ensure participants are competent and have sufficient underpinning knowledge of microscopy or, where necessary, remedial training in the use of microscopes at the TLM level.

Participants should understand the criteria for selection of microscopic features used by their organisation to examine hairs at the TLM level. Chapter 4 of this book and Chapter 2 from Robertson, J., 1999, *Forensic Examination of Hair*, Taylor & Francis Group, provides a suitable level of theory and underpinning knowledge for this module.

In addition to theory dealing with the microscopic examination of hairs at the TLM level, participants should understand what can properly be concluded and what cannot be properly concluded from TLM examinations. This includes understanding of type 1 and type 2 errors in the contemporary management of hair examinations/cases and the limitations of hair examination at the TLM level of examination. Content should also include detailed consideration of how to interpret hair findings at the TLM level and writing reports.

As examinations at the TLM level can contribute to a non-exclusionary conclusion, it is critical that hair examiners understand where and when hair examinations have contributed to wrongful convictions. Content should include consideration of case examples involving hairs where they have contributed to wrongful convictions. These may include both local and international cases. Some examples would include the Driskell and the Guy Paul Morin cases from Canada and the Splatt case from Australia.

These topics may be split between modules 3 and 4 where module 4 deals with the outcomes following comparison microscopy.

7.5.4.2 Competency Evaluation

Participants should be tested on their theoretical knowledge through short questions/quizzes and by delivering oral or written presentations on selected aspects of microscopic features designed to deepen the knowledge base of participants.

Participants should be tested on their practical skills through structured practical exercises.

Feature recognition can be tested with the use of authenticated images of microscopic features but must also include the examination of a range of hairs selected to show the range of microscopic features seen at TLM. The use of images alone is not acceptable as it does not test the ability of a trainee to use TLM effectively to fully capture the features in hair.

Participants should classify and record features using a checklist and written descriptions as appropriate. Whilst there is no minimum or maximum number of hairs that should be examined, there should be sufficient samples to provide confidence that participants demonstrate genuine competence through demonstrating a very high degree of success in accurately classifying and recording the potential information available for each hair.

Practical exercises should be designed to build in complexity and proficiency tests should be designed to test competence from simple to more complex scenarios.

Note, in our view it is critical that trainees are exposed to a very large number of hair samples and on a frequent basis over a sustained period of time. Visual literacy develops through exposure. The examination of hairs at the TLM level requires significant personal and organisational commitment and is not a part time activity to be practiced on an occasional basis.

The capstone exercise for this module should be a series of mock case scenarios of increasing complexity. Participants should demonstrate that they can reach the correct conclusion for each scenario and write a suitable report for each scenario demonstrating that they understand the limitations of hair examination at the TLM level.

7.5.5 Module 4

Develop knowledge of and competency in the use of comparison microscopy

On completion of this module participants will be able to:

- Demonstrate an understanding of, and practical skills in the use of, a comparison microscope.
- Demonstrate an understanding of the comparison process for hairs including pattern recognition.
- Demonstrate the ability to correctly exclude questioned/recovered hairs from known samples at the TLM level.
- Demonstrate an understanding of how to interpret the outcomes of hair comparisons where hairs have not been excluded.
- Write reports on the outcomes of hair comparisons.

7.5.5.1 Content

As this is the final technical module for level 3 training it is assumed that participants will have successfully completed level 1 and 2 training and modules 1–3 of level 3 training.

New theory and knowledge for this module should include the knowledge of comparison microscopy and especially the importance of achieving light balance. As a high-end comparison microscope may be more complex than routine TLM this may require some additional microscopy theory.

Building on previous modules, the content in this module should pay particular attention to variation in microscopic features and pattern recognition within hairs and between hairs.

The outcome of a comparison microscopic examination may result in an exclusion or a non-exclusion. In the latter context it is critical that examiners understand type 1 and type 2 errors and how these influence interpretation and reporting. Module 5 deals with interpretation and reporting in greater detail, hence, in this module content may be limited to an introduction to the considerations arising from the outcomes of hair comparisons and laying the foundation for a more detailed consideration of this topic in module 5.

7.5.5.2 Competency Evaluation

Participants should demonstrate competence in the use of a comparison microscope and an understanding of underlying microscopy theory. This may involve some testing of theoretical knowledge through short questions/quizzes or an oral test.

Practical competence should be tested through the use of a structured series of tests of increasing complexity with a range of expected outcomes. Participants should demonstrate their ability to assess and compare complex patterns of microscopic features of hair.

The capstone exercise for this module should be a series of mock case scenarios of mixed complexity. Evaluation of participants should not be restricted only to the final decision but include participants being able to logically explain the basis for technical conclusions and interpretation from an evidence perspective. Whilst some tests may have simple and definitive outcomes more complex tests may have a range of acceptable outcomes.

Participants should prepare written reports for a range of outcomes including non-exclusions. Participants should be tested in their ability to communicate such outcomes to investigators.

In module 5, exercises completed in this module may form the basis of competency evaluation such as giving evidence.

7.4.6 Module 5

Develop knowledge of the interpretation and reporting of hair outcomes from TLM and comparison microscopy.

On completion of this module participants will be able to:

- Understand the factors that impact on the interpretation of hair findings at the TLM/comparison microscope level.
- Understand contemporary thinking on the use of evaluative reporting as it applies to hair findings.
- Understand contemporary thinking on the use of verbal scales as it applies to hair findings.
- Demonstrate the ability to write reports for a range of outcomes from hair findings at a TLM/comparison microscope level.
- Understand the obligations as a professional and as an expert witness in giving hair evidence.
- Demonstrate high-level oral communications skills in accurately conveying hair findings to a range of stakeholders.

7.4.6.1 Content

This module may be delivered at the same time as later modules in level 3 training although the capstone exercise should be at the end of the complete training programme. This module is built on the underlying knowledge from all previous modules. Chapters 5 and 6 of this book provide a sound theory basis for this module. Although somewhat dated Chapter 7 by Gaudette in Robertson, J., 1999, *Forensic Examination of Hairs*, Taylor & Francis Group, also provides useful background and information.

AS 5338 parts 3 and 4 deal with evidence interpretation and reporting and provide useful generic guidance.

It is assumed that organisations will have in place a specific training programme in delivering expert evidence. Organisations may also have specific training in the interpretation of scientific findings. Training delivered as part of this module should be incorporated or adapted to compliment more generic training but should address the specific aspects relevant to hair examinations and findings.

7.4.6.2 Competency Evaluation

We recommend that training in this module is delivered largely in a tutorial/seminar mode where participants would be expected to have completed pre-reading and be positioned to discuss and debate demonstrating a deep knowledge of the topics being discussed. This may be complemented with written exercises where participants would be expected to conduct independent research and present their findings.

Included in a tutorial setting should be sessions in which participants explore the reasons underlying wrongful convictions where forensic science, and if possible, hairs, have been a contributing factor. Given the importance placed by some of the historical issue within the hair group of the Federal Bureau of Investigation (FBI), participants should demonstrate a knowledge and understanding of the root causes of issues with testimony and reporting identified by the FBI (Anon, 2018b).

Drawing on the case scenarios participants should prepare reports and present their findings to a range of stakeholders to demonstrate their communications skills.

Participants should present findings from a number of case scenarios as evidence in a mock court situation. Where possible participants should attend a number of real court situations and observe experienced expert witnesses giving evidence.

CHAPTER 8

Acquired Characteristics

Adine Boehme (Guest author)

8.1 INTRODUCTION

The introduction and progression of forensic DNA analyses (nuclear [nu] and mitochondrial [mt]) over the past 35–40 years has largely replaced the routine practice of forensic hair examination as a source discernment tool. However, the purpose of hair examination does not only seek to allocate hairs to the donor. Forensic hair examination has the power to provide contextual information relating to the "how" as well as the classic question of "who". This is a strength within forensic hair examination that is an acknowledged weakness of its DNA analysis counterparts, noting the two techniques reciprocal weaknesses and strengths, it is no surprise they are routinely paired.

Hairs gather and accumulate features along their journey of growth and the exposures they are subject to. These features are referred to as acquired characteristics. The forensic hair examiner is skilled in identifying and differentiating features acquired naturally as part of a regular hair life cycle (attributed to daily living) and those acquired unnaturally. Namely, as a result of violent force or unfortunate events which have resulted in a police investigation.

The identification of acquired characteristics may occur at any stage of the examination (either level 1, 2 or 3—see Chapter 4.1); however, it is generally during level 2 and 3 examinations that hairs present anomalies that capture the examiners attention.

The acquisition of hair characteristics can be thought of as analogous to the skin on a person's hands. During the life of a pair of hands the skin acquires wrinkles, callouses, freckles and scars that are telling of their journey and activities. Hair shafts also acquire features. Commonly observed features acquired by natural means can include evidence of cutting, bleaching, colouring, permanent waving, straightening, shaving, thickening, thinning and other treatments as per personal taste.

DOI: 10.4324/9781315210650-8

Hair features are also unknowingly acquired through environmental exposure; for example, sun bleaching, broken/split hairs, health of the individual, personal hair care habits and general wear and tear of brushing, combing and styling. These naturally acquired features are identified and recorded during the microscopic examination (see Appendix 4.2a) process as they are highly variable across different individuals.

Hairs also acquire features at an elemental level, where the elemental profiles resulting from specialised chemical tests can indicate history of exposure to drugs (illicit or prescribed), poisons, toxins or time spent in particular geographic regions. Elemental level features are outside the remit of microscopic hair examiner, but the microscopist may consider a referral to a forensic chemist if the investigation warrants.

Finally, it is the features that lie outside of the normal range of natural acquisition, or unnaturally acquired features are the interest of this chapter. Presented here are a number of scenarios from events involving violent force or misfortune, concentrating on the microscopic features observed in each event. These observations may be a point of comparison for future forensic hair examinations possibly projecting context.

It should be noted that the scenarios and their accompanying images are presented to promote understanding of the presence and variation of damage and how it relates to the particular scenario. It is not the intent of these scenarios to represent an exhaustive list of possibilities.

The examiner is encouraged to compile a reference collection or library of examples of acquired damage from known/controlled events. For example, samples of hair from a hairdresser cut using scissors, razors and clippers. The reference library should include detailed notes, observations and images. This allows the comparison of specific or "signature" cuts observed within casework to be compared with the "known" examples in the library. This may allow the examiner to exclude certain types of cuts from being the source of the damage, or more probatively narrow down the list of possible causes.

The image library may also be expanded by including damage brought about by controlled simulation experiments, or taken from casework where the damage to the hair is from a known occurrence.

8.2 EXAMINATIONS WITH ACQUIRED CHARACTERISTICS

Recording detailed descriptions of examinations and observations (as with all forensic hair examinations) is critical. Observations of acquired damage may be observed during any level of examination and should be recorded contemporaneously. The examiner may not be aware of their contextual relevance at the time, but making note of the presence or absence of damage is key.

TABLE 8.1   Acquired Damage Checklist

Damage Observed	Exhibit ID: 123	Exhibit ID: 456	Exhibit ID: 789
Location	Mid shaft (position marked with * on level 1 exam form, Appendix 4.2a)	Entire shaft	Root ends
Damage observation/ description	Sharp angular slice/cut marks (see Image A1)	Bubbled medulla (see Image A3)	Hair clump examined; all root types are anagen.
Debris	Numerous glass fragments in exhibit bag	Ash (black coloured debris with smoky odour) in envelope	
Contaminants	Apparent fungal growth on cuticle	Fine debris on cuticle	Apparent blood visible at proximal end, under 40× stereo. (Hemastix™)
Comments	Hairs appear wet		Hairs received as a ponytail like clump, assumed to be the same source.

A more detailed acquired damage examination may become relevant at a later stage, and may be revisited in subsequent examinations. An example—Table 8.1—provides a checklist of acquired damage. A systematic examination may include; notes, sketches/annotations and images. Each type can serve to answer different questions and are all relevant to capture the damage holistically.

Completing a checklist similar to the one above for each hair may reveal damage trends across a series of questioned hairs. On review of case notes, it may be observed that all of the hairs have mid shaft angular slice marks and fungal growth. Observations may also be made at a gross exhibit level. For example, where an entire clump of hair is submitted, and on initial inspection, all roots appear anagen; or at an individual hair level.

8.3 REPORTING ACQUIRED CHARACTERISTICS

As with all forensic hair examination reports, it is important to include reference to any comparisons undertaken, conclusions drawn and limitations associated with the respective examinations and conclusions.

Conclusions relating to acquired damage should reference that they are drawn from comparisons, should be specific and reflect the work that was undertaken, with sufficient caveats indicating that the conclusions are not to the exclusion of all other possibilities.

8.4 NATURAL AND UNNATURAL HAIR LOSS

The reality of natural hair loss has been previously discussed (see Chapter 2.4), Brooks and Robertson (2012), where the majority of hairs collected at crime scenes/from persons or items of evidence are telogen stage hairs—those that are lost naturally by some tens or hundreds per day. Unnatural hair loss in the forensic context presents entirely differently. Hair evidence with anagen roots and hairs with no roots are equally of interest. First, the presence of anagen roots in a number of hairs immediately indicates that the hair was lost unnaturally—whether it is scalp hair or body hair there is some force required to remove actively growing hairs from their follicles. Second, hairs without root ends means that the hair became detached from the individual somehow and was recovered by the crime scene scientists. The obvious question is "how" then "why" the hair has no root ends. The aim of this chapter is to reflect on some case work examples with specific characteristics in instances where cause and effect meant the hair was lost unnaturally. Our intention is to give a broad overview of each case using light micrographs to illustrate the acquired features and how they pertain to the facts of the investigation.

8.5 CUT HAIRS—PROFESSIONAL HAIRDRESSING

A level 2 or 3 hair examination observes and records general features of evidentiary hairs (see Appendices 4.2a and 4.2b) including shaft profile, colour, root types and tip types. Scalp hairs commonly have cut tip ends that may be blunt cut or cut at an angle depending on the cutting tool used and how recently the hairdressing was undertaken; rounded tips indicate a period of time between cuts. Figure 8.1 shows the difference between a freshly cut hair tip and a hair tip uncut for six months. Important to note here is the implement used for the hairdressing as each tool tends to have a "signature" cut that shapes the hair when freshly cut. A study of hair tip ends using different cutting tools was undertaken by Brough (2007). Figure 8.2 shows four scanning electron micrographs of four different cutting tools and their "signature" cut edge on a hair tip.

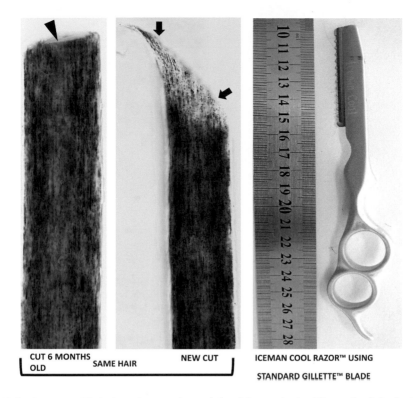

CUT 6 MONTHS OLD | SAME HAIR | NEW CUT | ICEMAN COOL RAZOR™ USING STANDARD GILLETTE™ BLADE

FIGURE 8.1   Hairdressing tools and freshly cut hair. Tip end of the hair cut six months previously is rounded on the edges (arrowhead) from weathering, whilst the new cut is angled and sharp (arrowheads). The razor is a typical hairdressing tool.

8.6 MOTOR VEHICLE COLLISION AND WINDSCREEN IMPACT

A man walking on the side of the road at night was deliberately hit from behind. His body was removed from the scene into the back of a utility vehicle and dumped at a different location. The recovered scalp hairs were mainly from the tray back of the vehicle. Blood and glass fragments were retrieved from the hair samples. Relevance here was that the recovered hairs had no root ends despite the relatively abundant sample, implying that these hairs were lost unnaturally. At the time of the examination much of the contextual information was withheld (the hit and run and relocation of the body). Low-power microscopic examination (LPM) of both mounted

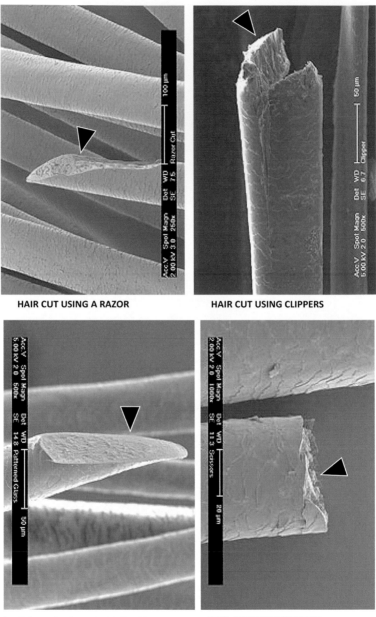

HAIR CUT USING A RAZOR   **HAIR CUT USING CLIPPERS**

HAIR CUT USING THE FB1™ GLASS BLADE   **HAIR CUT USING SCISSORS**

FIGURE 8.2 Scanning electron micrographs of tool signature cuts. (a) and (b) The scanning electron micrographs (Brough, 2007) show hair tip ends cut using a razor (arrowhead) and clippers (arrowhead). **(c) and (d)** The scanning electron micrographs (Brough, 2007) show hair tip ends cut using the FB1™* (arrowhead) and scissors (arrowhead). (*The FB1™ is a glass blade for cutting hair developed by Frank Bisson, 2007.)

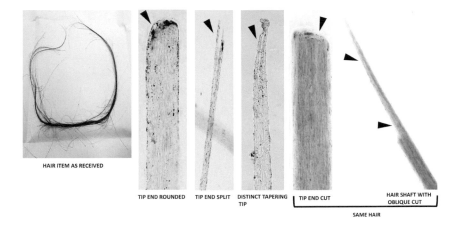

HAIR ITEM AS RECEIVED

TIP END ROUNDED   TIP END SPLIT   DISTINCT TAPERING   TIP END CUT   HAIR SHAFT WITH
                                            TIP   OBLIQUE CUT

SAME HAIR

FIGURE 8.3 Compares expected hair tip ends and unexpected mid shaft cuts. The hair item as received had no root ends and appeared to be cut mid shaft. These hairs displayed the variation of tip ends expected including rounded tip ends, split tip ends, distinct tapers and cut ends. However, the unexpected oblique cuts found in the shafts of many of these hairs indicated that the hairs were lost unnaturally and this particular cut may be related to the incident.

and unmounted samples indicated the hairs shafts were without root ends, tip ends were generally a combination of natural tapers/rounded tips and damaged/split. Hair shafts were cut at various lengths along the shaft, but the absence of a known sample leaves overall length as conjecture. Of particular note though were the shape, consistency and frequency of the mid shaft cuts. Figures 8.3 and 8.4 illustrate this case and the cuts to the hair shaft once the context of the crime was relayed could be attributed to glass cuts from the victim's head impacting the windscreen of the vehicle.

8.7 BLUNT-BLADED IMPLEMENT

A drug-related/-induced attack with a blunt-bladed meat cleaver resulted in the death of two people in their home. A fire was lit attempting to destroy the evidence, but the fire failed to burn sufficiently well to achieve the aim. Clumps of long human scalp hair from the female victim were located throughout the scene, coated in blood and ash. None of these hair clumps retained any root ends. Figure 8.5 shows one such hair clump. Examination of the hair clumps, mounted individual hairs and hair samples were recovered from the blade of the cleaver and were compared with known samples and all were attributed to the

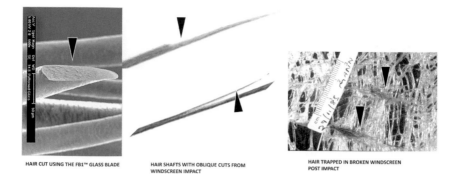

HAIR CUT USING THE FB1™ GLASS BLADE   HAIR SHAFTS WITH OBLIQUE CUTS FROM WINDSCREEN IMPACT   HAIR TRAPPED IN BROKEN WINDSCREEN POST IMPACT

FIGURE 8.4 Compares hairs cut with glass blade and windscreen impact. The scanning electron micrograph illustrates a single hair cut, whilst the FB1™ glass blade indicates an oblique angled clean cut with almost no cortical disruption (Brough, 2007). Also shown is a broken windscreen with human scalp hair trapped between the breaks (arrowheads)—hair that would also display clean, oblique cuts, as shown in the other images. The images of the hair shafts recovered after impacting with the windscreen also provide examples of hairs cut with glass that appear both oblique and cleanly cut as the hair deliberately cut during hair dressing with the FB1™ glass blade. The latter are acquired characteristics that link the hair of the victim to the windscreen impact that was ultimately the cause of death.

female victim. The cut hair clumps examinations revealed interesting acquired features. As noted, the recovered hairs tended to be in clumps, had no root ends and tip ends that had been previously cut then rounded through wear. The hair had been artificially coloured numerous times and given the length of the known hair (that had roots) the hair clumps appeared to have been cut off within several centimetres from the scalp. Whilst obviously cut through the upper part of the hair shaft, these cuts were not clean cuts as from a sharp blade (see 8.8). The cut ends were often flattened/crushed and folded or apparently torn before the actual cut achieved that may indicate the cleaver was blunt and required a degree of mechanical force. Figure 8.5 also shows the crushed/folded cut hair end.

8.8 SHARP-BLADED IMPLEMENT—KITCHEN KNIFE

A visit between neighbours, fuelled by excess alcohol, led to a disagreement and one male neighbour using a kitchen knife to stab the other in the chest and groin region. There was no dispute as to the weapon used

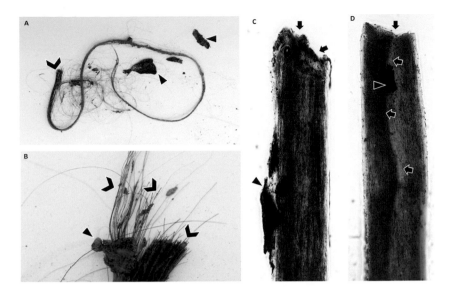

FIGURE 8.5 Abnormal cuts in hair shaft—blunt meat cleaver. The images/light micrographs (a) clump of hair as collected from the crime scene (a & b) and two more detailed images (c & d) of acquired damage features presumably from the meat cleaver. (a) Sample of hair evidence as received in the laboratory. The chevron indicates the "cut" end of the clump of hair, whilst the arrowheads indicate the ash and debris attached to the hair evidence. (b) Same sample as (a) but showing the cut end of the hair clump. Chevrons indicate the different hair lengths resulting from the cut/chopping of the hair. The uneven lengths may be the result of the bluntness of the meat cleaver where a clean cut was not possible causing the blade to cut then slide along the rest of the hair before finally cutting through. Arrowheads indicate ash and debris. (c & d) Arrows show the "chopped/flattened" hair ends, arrowheads indicate ash and debris attached to the hair shafts, and the white outlined arrows show a "crease" in the hair where the shaft has been flattened before cutting was achieved.

nor the two men involved. Hairs were recovered from the knife and the victim's body, and subsequently known samples from both men. In this case, the acquired features seen in the hair shafts (again no root ends were observed in the evidentiary hairs) are different to the glass cut hairs from the windscreen (Figures 8.3 and 8.4) and the crushed torn hairs from the blunt cleaver (Figure 8.5). The cuts from the kitchen knife on the shafts of the pubic region hairs present as ragged

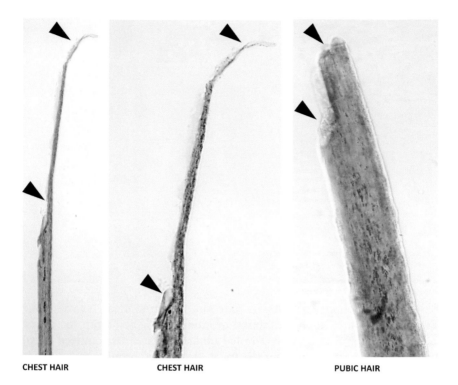

CHEST HAIR   **CHEST HAIR**   **PUBIC HAIR**

FIGURE 8.6   Hair shaft cut with kitchen knife. As previously described in Figures 8.1 and 8.2 cutting implements tend to have a "tool signature" cut. As opposed to the glass blade the cuts to these hairs are torn and ragged (arrowheads) tapering off to a wisp of cortex at the hair end (arrowheads). Similarly, as with the other hairs discussed previously, these hairs had no roots and the cut was affected mid shaft. The first two hairs were collected from the victim's chest and the third hair from the pubic region, the latter looking more broken and blunter (arrowheads) than the chest hairs.

and torn and taper off to a thin layer of cuticle/cortical cells. The chest hair appears broken rather than cut, see Figure 8.6.

8.9 BLUNT FORCE IMPACT— BROKEN/CRUSHED HAIR

Broken or crushed hair generally results from blunt trauma type impacts, specifically to the head or body regions. As acquired characteristics the breakage or crushing is usually noticed as disrupted and exposed cortical and cuticle regions of the hair shaft. Both elements

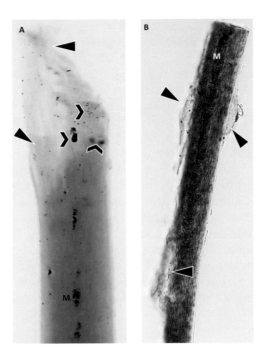

FIGURE 8.7 Crushed and broken hair shafts. Crushing or breakage of hair results in the exposure of the cortex/cortical cells and breaches the integrity of the hair. Once breached the softer cortex is easily further damaged from water, environmental effects, grooming and other pathogenic attacks. (a) Shows a hair that has been crushed mid shaft where the cortical cells are torn (arrowheads) away from the central core of the hair. Debris possibly from the implement located on the hair surface (chevrons) and M indicates an intermittent medulla. (b) The shaft of this hair has been broken (arrowhead) causing the cuticle and some attached cortex to lift away from the hair shaft (arrowheads). A medulla (M) is observed.

lack their former integrity and depending on the trauma inflicted the crushed or broken hair ends may appear both ragged and embedded with debris, see Figure 8.7.

8.10 BLUNT FORCE IMPACT—FLOOR SAFE DOOR

A cold case was revisited after new evidence was tendered. The context given to the examiner was limited to the proposed scenario of one partner suspected of murdering the other. The hair evidence was presented in a large plastic bag containing numerous reddish coloured hairs. The recovered hairs were from the victim and most had been collected from the study floor and waste paper bin. LPM examination of all recovered hairs indicated that apart from two (2) hairs with telogen root types no other hair had a root end. The two hairs with the telogen roots were a darker brown and could not have come from the same source as the others in the wastepaper basket/study floor. Distinctive feature confined to the reddish hairs was the brown debris embedded into the apparent crushed mid-shaft breakages, see Figure 8.8. The diligence of the crime scene examiner noting this detail contemporaneously was crucial to the re-investigation, as with context it turns out that the victim (who had the reddish dyed hair) had been hit over the head with a floor safe lid then shot. The debris imbedded within the shafts of the reddish hairs was likely to be from the safe door. Mitochondrial analysis of the two dark brown hairs with telogen roots revealed these did come from the suspect (known hair samples obtained during re-investigation) but as the bulk of the hair evidence was collected at the time of the murder before which the suspect and victim were living together the suspect's hair found amongst the victim's hair was ruled likely and therefore no further hair case to answer.

8.11 ARSON/FIRE VERSUS COSMETIC HEAT STYLING

Burnt human hair exhibits a number of changes including colour change, singeing, expansion and bubbling. Burnt hair when viewed at LPM, and more so at TLM, is unmistakable. However, for it to be of evidentiary value the effect of applying hairstyling tools to the hair needs to be considered. Such tools include straightening irons, heated rollers, blow drying, hot pressing combs, some of which become hot enough to burn and blister the skin if touched. Severe burning and scalding occur in liquids at 140°C within 5 seconds (American Burn Association (2000), Scalds: A Burning Issue), and heated hair styling tools claim to achieve temperatures 100–220°C (Applica Consumer Products, 2005).

Pangerl and Igowsky (2007) conducted a study of human head hairs exposed to heat within a forensic context. They found, briefly, that exposure to an enclosed heated environment (furnace) for a maximum of 20 seconds showed some discoloration of the hairs at as low as 175°C and bubbling at 200°C. Changes along the entire length of the hair occurred at 300°C. In a smaller system meant to simulate a heated styling tool the majority of hairs showed colour changes and bubbling when exposed to testing temperatures of 235°C and 250°C. The relevance of this information relates to the case described in Section 8.12.

FIGURE 8.8 Crushed hair from floor safe lid. The light micrographs here show two ends of the same hair recovered at the crime scene of this murder. Hair A has a cut tip end that is slightly rounded (arrow) and many cortical fusi (CF) can be observed throughout the shaft. Hair B shows the crushed/broken mid shaft closer to the root end where the cortical cells (arrowheads) are exposed and torn. The chevrons indicate surface debris possibly from the safe lid used as a weapon causing the crushing breakage of the hair shafts.

8.12 ARSON/EXPLOSION

A man thought to be connected with an act of arson causing a major explosion was located interstate, injured and requiring hospitalisation. From the footwell of his vehicle numerous hairs were recovered, along with an electric razor, both of which were submitted for forensic examination. It was thought that the suspect attempted to change his appearance to avoid capture. None of the examined hairs had root

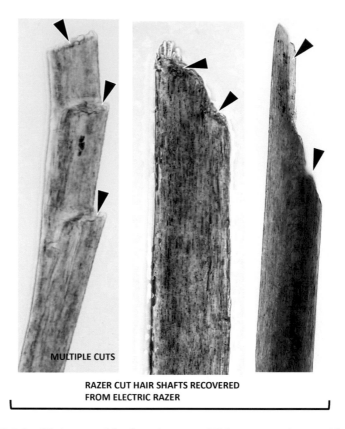

MULTIPLE CUTS

**RAZER CUT HAIR SHAFTS RECOVERED
FROM ELECTRIC RAZER**

FIGURE 8.9   Hair cut with electric razor. This case study provides hair evidence that links both the suspect to the hair and the razor used to remove it and the suspect to the crime scene in another state. The evidence chain involves the suspect shaving off his scalp hair with an electric razor (both hair and razor were recovered from the suspect's vehicle). The removal of the hair relates to the burnt/heat affected state of the suspects hair that resulted from an arson/explosion attempt on a house in another state's city. In this figure, the shaved hair evidence indicates the cuts (arrowheads) in the hair shafts from the electric razor. None of the recovered hairs had root ends. Figure 8.2 part 1 illustrates scanning electron micrographs of a hair cut with a razor where the appearance of the two sets of razored hairs are quite similar.

ends and a transmitted light microscopy (TLM) comparison of the hair tip ends with some other hairs recovered from the electric razor cutting blades indicated that the hair had been shaved from the head using the razor submitted—see Figure 8.9. Many of the hair tip ends ranged from distinct tapers to rounded in shape suggesting previous

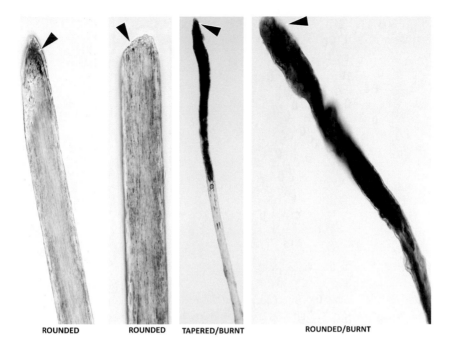

ROUNDED   ROUNDED   TAPERED/BURNT   ROUNDED/BURNT

FIGURE 8.10 Tip ends from both normal and heat affected hair. Illustrated here in these light micrographs are the tip ends of four hairs recovered from the suspect's vehicle. Two of the hairs have typical rounded tips (arrowheads) generally observed in hairs that have been cut for some months or longer. The other two hairs have burnt tip ends, one being a natural tapering (arrowhead) tip and the other, possibly originally a rounded tip now burnt and swelling (arrowhead).

hair cutting, see Figure 8.10. Of note though were the many hairs that had the features indicative of exposure to heat. Such features included colour changes, singed ends, bubbling and other distortions of the hair tip/and or shafts—see Figure 8.11. As suggested above by Pangerl and Igowsky (2007) even a 20-second exposure to high temperatures resulted in significant and recognisable "acquired" features of the hair—Figure 8.11.

8.13 EXPLOSION

We received a request to examine hair collected from the body of a victim of a suicide bombing. None of the authors here had previously examined hairs that had been specifically recovered from an explosion.

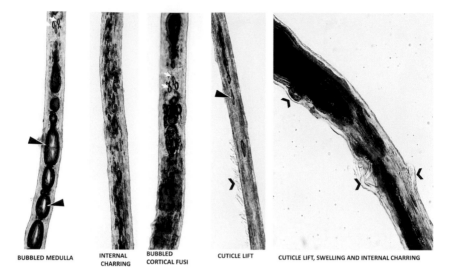

BUBBLED MEDULLA   INTERNAL   BUBBLED   CUTICLE LIFT   CUTICLE LIFT, SWELLING AND INTERNAL CHARRING
                   CHARRING   CORTICAL FUSI

FIGURE 8.11   Features seen in heat affected hair. The final figure associated with Section 8.5 shows a number of features associated with heat exposure and burning of hair. The intensity of the heat and its duration are factors in the features seen within and on the hairs surfaces. Pangerl and Igowsky (2007) described discolouration, bubbling, shaft swelling in their study. Illustrated here are bubbling within the medulla (arrowheads); bubbling within the cortical fusi (white double arrowheads); internal discolouration/charring (arrows); cuticle lift (chevrons) and burnt hair end (double arrowheads).

The submitted hairs were either loose hairs or embedded in scalp tissue. A number of the loose hairs were mounted for both LPM and TLM and extensive micrography undertaken. There were a number of anagen stage roots where both LPM and TLM examination was undertaken and the resulting images are seen in Figures 8.12–8.14, observations show what we assume are the damaging effects causing the acquired characteristics resulting from explosive forces on the victim's hair. In anagen hairs with sheath material debris and what appeared to be glass fragments were embedded in the sheath. The cuticle layer in many hairs was peeled away from the cortex (Figure 8.13a and b) and in some hairs the cuticle was entirely removed with the cortex fully exposed and unprotected—Figure 8.14. In other hairs the cuticle was cracked along one side of the hair shaft (Figure 8.13c and d), whilst some hairs appeared sliced with a thin region of cuticle and cortex remaining.

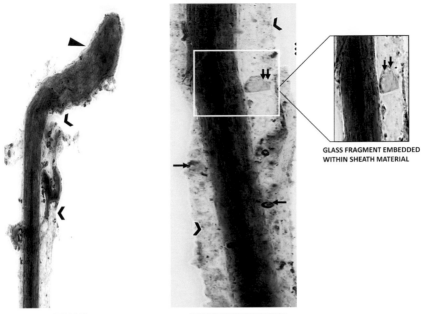

ANAGEN ROOT   ANAGEN ROOT WITH SHEATH

GLASS FRAGMENT EMBEDDED
WITHIN SHEATH MATERIAL

FIGURE 8.12 Anagen roots post bomb blast. The light micrographs here show a single hair with an anagen root. This hair amongst others was recovered at the scene from the victim of a suicide bombing. Nuclear DNA results (analysed elsewhere) identified the victim as being the source of the recovered hairs. The anagen root (arrowhead) is obvious as is the sheath (chevrons) surrounding the hair that in this and other circumstances would indicate that this hair was lost unnaturally. An intact anagen hair amongst other anagen stage hairs immediately suggests that hair was forcibly removed. Undoubtedly force was applied here, and of note is the debris (arrows) observed on both the surface and within the sheath (chevrons) material. The double arrows show what appears to be a glass or other material fragment embedded in the sheath (chevrons) located next to the hair shaft.

8.14 HEAD LICE

Incidents of child abuse and neglect are seen throughout the course of police investigations. Sometimes indications of neglect/abuse can be seen as injuries, malnutrition or lack of hygiene, for example where an infestation of head lice is so rampant that the lice may move from the scalp to other parts of the body. There are specialist forensic entomologists/paediatricians who can determine the frequency of life cycles of

CUTICLE PEEELED AWAY FORM SHAFT   CUTICLE PEEELED AWAY FORM SHAFT

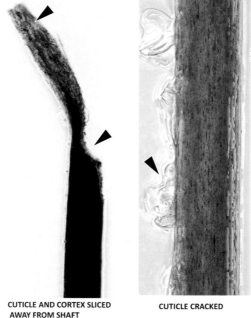

CUTICLE AND CORTEX SLICED   CUTICLE CRACKED
AWAY FROM SHAFT

FIGURE 8.13   Cuticle features post bomb blast. (a) and (b) Two different hairs recovered from the suicide bombing victim showing possible force effects on the hair cuticle (arrowheads). In both hairs the cuticle layer appears to have peeled away from the shaft exposing the inner cortex. The rest of the cuticle is intact in one hair, but in the other hair the cuticle appears disrupted. Debris (chevrons) is again present and in one hair a continuous medulla (M) is seen. (c) and (d) The cuticle with a layer of cortex attached appears to have been "sculpted" away from the hair shaft (arrowheads). The cause, though it is conjecture, may have resulted from sharp debris cutting the hair with force from the explosion. The other hair is a classic example of cracked cuticle (arrowhead) along the hair shaft—one side only. The cuticle on the other side of the hair is intact.

DISRUPTED SHAFT CORTEX ONLY – NO CUTICLE REMAINS

FIGURE 8.14 Cortical features post bomb blast. Finally, this light micrograph shows a hair tip end where the entire cuticle is missing—leaving the cortical cells completely exposed. Debris (chevrons) is present on the surface and within the hair cortex that is fragmenting along the edges (arrowheads). This is quite an extreme example of a very damaged/disrupted hair shaft.

insects including head lice, thereby aging the length of the infestation. When a scalp hair is literally covered in empty "nit" egg cases from the root to the tip end this is a severe infestation, see Figure 8.15. Whilst not actually an "acquired damage characteristic" of hair, a severe or even a mild infestation of head lice will be noticed by the hair examiner. Case dependent circumstances will determine further investigation.

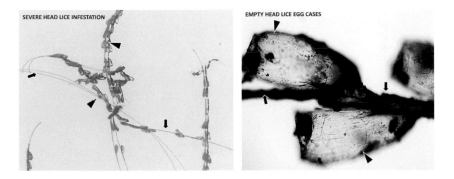

FIGURE 8.15 Head lice infestation. The top image shows head lice egg cases (arrowheads) clustered along individual hairs (arrows). The egg cases are glued to the hair by the adult louse, but as shown in the lower image these cases are empty. The hatched/adult lice move to the scalp to blood feed. This is a severe infestation where nearly all the hairs would appear as the two images shown here.

8.15 CONCLUSION

Hair acquires characteristics all the time indicating to some degree a life led. However, the state of the evidentiary hair as received can also indicate something other than "normal circumstances". Anomalous features such as a clump of hairs with anagen root or hairs with no root ends should be something of a red flag to the hair examiner warranting an examination record. The brief case scenarios and accompanying images described here show how careful examination and recording of evidential hairs can link the acquired damage/features seen in the hair evidence to the sequences of events that occurred during the commission of the crime being investigated.

Glossary

Alopecia: A general term for baldness or hair loss.

Alopecia areata: A form of alopecia in which hair is lost in one or more patches.

Alopecia diffusa: Diffuse, generalised loss of scalp hair which may result from malnutrition, a range of health issues or drug use.

Alopecia, male pattern: Also known as androgenic alopecia it is called "male pattern baldness" and results from the action of androgens and is most commonly seen in males as they age.

Alopecia traction: Hair loss resulting from strong and sustained or repeated tension or traction on the hair follicles—may result from some hair styling or treatments.

Amorphous medulla: A medulla that has no microscopically visible distinct form, pattern or shape.

Anagen: The active growth phase of a hair follicle in the hair growth cycle (Greek: *aner* = up, *genesis* = mode of formation).

Androgens: Steroid hormones, including testosterone and androstenedione, which are required for the normal development of secondary sexual hair and the initiation of common baldness.

Apocrine gland: A type of sweat gland which forms in some follicles.

Arrector pili muscle: Smooth muscle attached to the posterior side of the hair follicle and which runs up at an angle to the papillary layer of the dermis. Contraction of the muscles due to cold or fright produces "gooseflesh".

Buckling: An abrupt change in shape and orientation of the hair shaft often with a slight twist, often seen in public hairs.

Bulge: A bulge which develops on the side of the hair follicle during development and which becomes the site of attachment for the arrector pili muscle. It marks the lowest point of the permanent part of the follicle and contains stem cells which divide to start the next anagen phase.

Canities: The term used for greying or whitening of hair.

Catagen: The transitional phase of the hair follicle from the active anagen growth phase to the resting telogen phase of the hair growth cycle (Greek: *cata* = down, genesis = mode of formation).

Caucasoid: A term used in anthropology to denote a major group of human beings of European origin.

Cell membrane complex: CMC are intercellular contacts that cement the cortical cells together and contribute to the tensile strength of the hair.

Characteristic: A macroscopic or microscopic attribute or feature of a hair.

Class characteristic: Properties of items that can be associated only with a group and never with an individual source.

Colour: An aspect of objects that may be described in terms of hue, lightness and saturation. Hairs colour assessed at a visual and at a microscopic level may differ in perception.

Comparison: The examination of two or more hairs to evaluate whether they could have come from the same source.

Continuous medulla: A medulla where there are no interruptions or breaks along the length of the shaft of the hair.

Cortex: The anatomical region of the hair shaft forming the hair shaft with the outer boundary being the cuticle. The cortex comprises elongated or fusiform shaped cells. There may be a central defined anatomical region called the medulla present in the cortex.

Cortical fusi: Small spaces between cortical cells that may be filled with air or liquid and can be seen with transmitted light microscopy (TLM).

Cortical texture: The appearance of the cortex defined by how visible the margins of the cortical cells are when viewed with TLM. Cortical texture may be classified as smooth or coarse.

Cracked cuticle: A cuticle with linear breaks that is perpendicular to the length of the shaft.

Cross-sectional shape: The shape of a hair shaft cut and viewed at a right angle to its longitudinal axis.

Cuticle: The outermost region of the hair composed of layers of overlapping scales.

Cuticle thickness: The relative size of the cuticle from its outer margin to the cortex when viewed microscopically.

Cycle: The hair growth cycle is the repeated growing (anagen), transition (catagen) and resting (telogen) phases of the hair follicle.

Dendrite: A branching cytoplasmic projection of a cell.

Dermal papilla: A projection of dermal tissue into the base of the hair follicle bulb. It controls the physical characteristics of the hair as well as hair growth and is the location of the androgen receptor necessary for the development of sexual hair at puberty.

Dermis: The true skin. It lies beneath the epidermis and contains hair follicles, sebaceous glands, blood vessels, nerves, muscles and so forth. Its basic structure is collagen and other elastic tissue.

Distal end: The end of the hair away from the root.

Endocuticle: The layer on the inner side of the cuticle.

Epicuticle: A hydrophobic (water repellent) layer which surrounds each cuticle cell.

Epidermis: The outer protective layer of the skin.

Eumelanin: A type of pigment granule, ellipsoidal to oval in shape resulting in dark brown and black colours in hairs.

Exemplar: Synonymous for known (an item of know source) or a representative sample taken from a known source for comparison to a questioned item.

Exocuticle: The layer on the outer side of the cuticle cells.

Feature: Synonym for characteristic.

Follicle: An invagination of the epidermis which contains the root of the hair.

Follicular tag: Tissue from a hair follicle that can be attached to the root end of the hair which has been removed from the follicle.

Fungal tunnels: Air pockets in the hair shaft caused by fungal growth.

Fusiform: A term that refers to the spindle-shaped cells (tapered at each end) that form the cortex.

Germinative cells: The dividing cells in the hair bub which give rise to the various cellular layers of the follicle and the hair fibre— sometimes called bulb matrix cells.

Glabrous: Free from hair.

Hair: A fibrous outgrowth from the skin of mammals.

Imbricate: A scale pattern with edges that overlap giving the appearance of "tiles on a roof" and giving a wavy pattern typically found in human hair.

Inconclusive: A conclusion that may be reached by a hair examiner due to the inability to exclude or include a questioned hair when comparing hairs with known hairs.

Inner root sheath (IRS): A sheath of three layers of cells (cuticle, Huxley layer, Henle layer) which surrounds the hair follicle and contributes to the shaping of the hair shaft and to holding the hair in the follicle during the anagen growth phase.

Item: Physical thing that is collected for forensic purposes.

Intermediate filaments (IF's): Microfibrils of keratin protein which combine to form the keratin macrofibrils that fill the cortical cells of the hair fibre.

Interrupted medulla: Where the medulla is not continuous along the entire length of the hair shaft.

Keratin: A class of sulphur containing highly insoluble fibrous proteins found in hairs, nails, horns and feathers of animals.

Keratin-associated proteins (KAP's): A matrix of sulphur rich proteins in which the keratin intermediate filaments are embedded.

Known sample: A sample from a known source which may be required for comparison against a questioned sample; with respect to hair a sample taken from a known individual.

Lanugo: Soft, fine downy hairs which cover the foetus from about the fifth month of gestation and are usually replaced by vellus hairs shortly after birth. Lanugo hairs may remain for some time after birth.

Looped cuticle: Where the distal edges of the cuticular scales are curved from or cup towards the hair shaft.

Macroscopic: A feature or characteristic that is large enough to be seen without magnification.

Material: A general term including but not limited to "Physical material". Physical material may include any and all objects, gross or microscopic in size, living or inanimate, solid, liquid or gas, including the relationship between all such objects as they pertain to a crime.

Medulla: A central core of cells in the cortex of hairs. The appearance of the medulla is a useful feature to differentiate hairs of human origin from other animal hairs.

Medulla continuity: The continuous or discontinuous nature of the medulla from the proximal end to the distal end of the hair.

Medulla opacity: The appearance of the medulla as being either opaque or translucent when viewed with TLM.

Melanin: A natural pigment of which two forms eumelanin (dark brown to black) and phaeomelanin (reddish yellow) determine the colour of a hair.

Melanocytes: Dendritic cells in which the melanosomes are formed and delivered in the follicle to the developing hair.

Melanosomes: Granules that contain melanin pigment.

Microscopic: A feature or characteristic too small to be resolved by the unaided eye but large enough to be resolved when viewed with a microscope.

Monilethrix: A disorder of hairs that results in periodic nodes or beading along the length of the hair shaft with intervening tapering constrictions that are not medullated.

Outer root sheath (ORS): The outermost layer of the hair follicle, continuous with the layers of the epidermis.

Opaque medulla: A medulla with trapped air causing it to appear black when viewed with a microscope.

Ovoid bodies: Heavily pigmented roughly oval-shaped bodies seen in the cortex of some hairs.

Paper boat: Paper folded in such a way as to securely contain a small sample. Term used to describe a piece of paper that has been folded into thirds and then in from each side, sometimes referred to as a druggist's fold.

Phaeomelanin: Spherical pigments resulting in yellow and red colour in hairs.

Pilosebaceous unit: The combination of a hair follicle and a sebaceous gland.

Pigment aggregation: The appearance of pigment granules when they are concentrated in a mass that has a recognisable form such as steaked or clumped.

Pigment density: The relative abundance of pigment granules in the hair cortex when viewed with TLM.

Pigment distribution: The pattern of pigment granules in the cortex of the hair shaft.

Pigment granules: Small particles in the hair that impart colour to the hair as seen with a microscope.

Pili annulati: A hair disorder that results in a ringed or banded appearance with alternating bright and dark bands in the hair shaft.

Pili torti: A hair disorder that results in the hair shaft being flattened and twisted 180° numerous times along its axis. It is usually found at irregular intervals along the hair shaft.

Post-mortem banding: The appearance of an opaque band near the root end of hairs from a decomposing body. Hairs displaying this feature may or may not have visible roots attached.

Primary hair: The terminal hair that replaces lanugo hair at the scalp, eyebrows and eyelashes just prior or shortly after birth.

Proximal end: The portion of the hair towards the root.

Questioned hair: One or more hairs recovered during a forensic search procedure of unknown source.

Range: The variation of a specific feature or characteristic exhibited by a hair or hairs from one person.

Reference standard: With respect to hairs, a standard in a reference collection of hair to be used as a comparison item.

Representative sample: A collection of hairs from a specific body area that reflects the range of features or characteristics in a person's hair.

Root: The end of the hair shaft at the proximal end of the hair. The appearance of a human scalp hair root will depend on the growth stage of the hair.

Root sheath: The follicular tissue that holds anagen hairs in the follicle and is sometimes present near the root end in hairs that have been removed with some force.

Scales: Small plate like structures composed of keratin that together form the cuticle.

Scale pattern: The pattern formed by the edges of the scale (cuticle) cells on the surface of the hair shaft.

Sebaceous gland: Glands of the skin which may be attached to a follicle by a duct and produce sebum.

Serrated cuticle: A cuticle in which the outer margin has a notched appearance like a saw blade.

Sexual hair: Terminal hair which develops at puberty and replaces vellus hair at the pubis and axillae in both sexes and on the face of males. Also called secondary hair.

Shaft: The portion of a hair above the hair root and visually above the skin line.

Shaft form: The macroscopic shape of the hair.

Shaft thickness: The diameter of the hair shaft.

Somatic: An area of the body such as head, pubic or leg.

Splitting: Damage usually occurring at the distal end of the hair shaft when a hair splits or divides down the long axis.

Telogen: The final stage in the hair growth cycle when the hair has stopped growing and the root has assumed a club shape. Also called the resting phase (Greek: *telos* = end, *genesis* = mode of formation).

Terminal hairs: Long pigmented coarse hairs, sometimes with a medulla which are the final differentiation state for hairs at a particular site such as the scalp.

Testosterone: The principal male androgen hormone, necessary for the development of sexual hair in males and the development of common baldness.

Tip: The most distal end of the hair shaft. The tip end may be classified into a number of types based on the visual and microscopic appearance.

Translucent medulla: The appearance of a medulla when the medulla cells are filled with fluid and not air.

Trichohyalin: A major protein product of the medulla and inner root sheath cells. It is low in sulphur but contains citrulline and is highly cross linked.

Trichology: The study of hairs.

Trichonodosis: A condition of hairs characterised by apparent or actual knotting of the hair shaft.

Trichoptilosisis: A disease condition of hairs characterised by longitudinal splitting or fraying of the hair shaft.

Trichorrhexis invaginiti: A genetic disease condition of hairs characterised by a segment of bulbous, dilated hair enfolded into a concave hair terminal, recalling the appearance of a bamboo node. When broken the hair has the appearance of a golf tee.

Trichorrhexis nodosa: A condition characterised by the formation of nodes. The hair is weaker at the node and subject to breakage.

Trichoschisis: A condition of hairs characterised by a brittle hair with a transverse crack or a clean break.

Undulation: Change in true diameter along the length of the hair shaft that results in a change in cross sectional shape. This can give the hair a wavy appearance.

Vellus: Soft, fine unmedullated and unpigmented hair which replace lanugo hair before or shortly after birth.

Weathering: The effect of the environment on the hair shaft; wear and tear that cause damage to the cuticle and/cortex.

References

Adachi, K & Uno, H 1969, 'Some metabolic profiles of human hair follicles', in Montagna, W & Dobson, L (eds), *Advances in Biology of Skin. Hair Growth*, Volume 9, Oxford: Pergamon, pp. 511–534.

Adya, KA, Inamadar, AC, Palit, A, Shivanna, R & Deshmukh, NS 2011, 'Light microscopy of the hair: a simple tool to "untangle" hair disorders', *International Journal of Trichology*, vol. 3, pp. 46–56.

Aitken, CGG & Robertson, J 1986, 'The value of microscopic features in the examination of human head hairs: statistical analysis,' *Journal of Forensic Sciences*, vol. 31, pp. 546–562.

Aitken, CGG & Robertson, J 1987, 'A contribution to the discussion of probabilities and human hair comparisons', *Journal of Forensic Sciences*, vol. 32, pp. 684–689.

American Burn Association 2000, Scalds: A Burning Issue. http://www.ameriburn.org/Preven/2000Prevention/Scald2000PrevetionKit.pdf

Anon 1985a, 'Preliminary report – committee on forensic hair comparison', *Crime Laboratory Digest*, vol. 12, pp. 50–59.

Anon 1985b, 'Proceedings of the international symposium in forensic hair comparisons. Washington, DC: US Department of Justice', www.ncjrs.gov/pdffiles1/Digitization/116592NCJRS, pdf 31.

Anon 2005, 'Forensic human hair examination guidelines' https://www.asteetrace.org.

Anon 2006, Driskell inquiry. 'Commission of inquiry into certain aspects of the trial and conviction of James DRISKELL', *Transcript of Proceedings*, vol. 24.

Anon 2009, '*Strengthening Forensic Science in the United States: A Path Forward*', The National Academies Press, p.156. https://www.nap.edu/catalog/12589/strengthening-forensic-sciencein-the-united-states-a-path-forward.

Anon 2013, R v Hay, 2013 SCC 61. Docket: 33536.

Anon 2015a, FBI/DoJ Microscopic hair analysis comparison review. https://www.fbi.gov/news/pressrel/press-releases/fbi-testimony-on-microscopic-hair-analysis-contained -errors-in-at-least-90-percent-of-cases-in-ongoing-review.

Anon 2015b, 'ENFSI guideline for evaluative reporting in forensic science', https: ensfsi.eu/documents.

Anon 2015c, 'ENSFI Best practice manual for the microscopic examina-
    tion and comparison of human and animal hair', https://enfsi.eu/
    documents/best -practice-manuals.
Anon 2016a, 'Report to the President. Forensic science in criminal
    courts: ensuring scientific validity of feature comparison methods',
    *Executive Office of the President. President's Council of Advisors on
    Science and Technology*, www.obamawhitehouse.gov>ostp>PCAST
Anon 2016b, 'Department of Justice Proposed Uniform Language
    for Testimony and Reports for the Forensic Hair Examination
    Discipline, 2016', www.justice.gov/dag/file/877736/, download and
    'Supporting Documentation for Department of Justice Proposed
    Uniform Language for Testimony and Reports for the Forensic Hair
    Examination Discipline, 2016', www.justice.gov/dag/file/877741/
    download.
Anon 2017, 'An introductory guide to evaluative reporting', www.nifs.
    org.au.
Anon 2018a, FBI/DoJ microscopic hair comparison analysis review.
    https://www.fbi.gov/services/laboratory/scientific-analysis/fbidoj-
    microscopic-hair-comparison-analysis-review.
Anon 2018b, ABS report: root and cultural cause analysis of report and
    testimony errors by FBI MHCA examiners. https://vault.fbi.gov/root-
    cause-analysis-of-microscopic-hair-comparison-analysis/root-cause-
    analysis-of-microscopic-hair-compariosn-analysis-part-01-of-01/
    view.
Applica Consumer Products (2005). useandcaremanuals.com. http://
    www.useandcaremanuals.com/pdf/MH5003,MH5004760.pdf.
Astore, IPL, Pecoraro, V & Pecoraro, EG 1979, 'The normal trichogram of
    pubic hair', *British Journal of Dermatology*, vol. 101, pp. 441–444.
Baltazard, T, Dhaille, F, Chaby, G & Lok, C 2017, 'Value of dermoscopy
    for the diagnosis of monilethrix', *Dermatology Online Journal*, vol.
    23, no. 7. Retrieved from https://escholarship.org/uc/item/9hf1p3xm
Barman, JM, Astore, I & Pecoraro, V 1969, 'The normal trichogram of
    people over 50 years but apparently not bald', in Montagna, W &
    Dobson, L (eds), *Advances in Biology of Skin. Hair Growth*, Volume
    9, Oxford: Pergamon, pp. 211–220.
Barnett, P & Ogle, R 1982, 'Probabilities and human hair comparison',
    Journal of Forensic Sciences, vol. 27, pp. 272–278.
Birbeck, MSC & Mercer, EH 1957, 'The electron microscopy of the
    human hair follicle. Part 1 introduction to the hair cortex', *Journal
    of Bio-physiology, Biochemistry and Cytology*, vol. 3, pp. 203–214.
Bisson, F 2007, 'Beyond the cutting edge', www.frankbisson.com/fb1_
    glass_blades.html.
Boccalatte, S & Jones, MR 2009, 'Hair', in *Trunk Books*, Volume 1,
    Sydney: Boccalatte Pty Ltd.

Boonen, T, Vits, K, Hoste, B & Hubrecht, F 2008, 'The visualisation and quantification of cell nuclei in telogen hair roots by fluorescence microscopy, as a pre-DNA analysis assessment', *Forensic Science International: Genetics Supplementary Series*, vol. 1, pp. 16–18.

Bourguignon L, Hoste, B, Boonen, T, Vits, K & Hubrecht, F 2008, 'A fluorescent microscopy–screening test for efficient STR–typing of telogen hair roots', *Forensic Science International: Genetics*, vol. 3, pp. 27–31.

Brandhagen, MD, Loreille, O & Irwin, JA 2018, 'Fragmented nuclear DNA is the predominant genetic material in human hair shafts', *Genes*, vol. 9, pp. 640. doi:10.3390/genes9120640.

Brooks, E, Cullen, M, Sztydna, J & Walsh, SJ 2010, 'Nuclear staining of telogen hair roots contributes to successful forensic nDNA analysis', *Australian Journal of Forensic Science*, vol. 42, pp. 115–122.

Brooks, E, Comber, B, McNaught, I & Robertson, J 2011, 'Digital imaging and image analysis applied to numerical applications in forensic hair examination', *Science and Justice*, vol. 51, pp. 28–37.

Brooks, E & Robertson, J 2012, 'Natural and unnatural hair loss in the forensic context', in Preedy, V (ed), *Handbook of Hair in Health and Disease*, The Netherlands: Wageningen Academic Publishers.

Brough, I 2007, 'Beyond the cutting edge', www.frankbisson.com/fb1_glass_blades.html

Brunner, I I & Triggs, B 2002, *Hair ID. An Interactive Tool for Identifying Australian Mammalian Hair*, Melbourne: CSIRO Publishing.

Chao, J, Newsom, AE, Wainwright, JM & Matthews, RA 1979, 'Comparison of the effects of some reactive chemicals on the proteins of whole hair, cuticle and cortex', *Journal of the Society of Cosmetic Chemists*, vol. 30, pp. 401–413.

Chase, HB 1965, 'Cycles and waves of hair growth', in Lyne, AG & Short, BF (eds), *Biology of the Skin and Hair Growth*, Sydney: Angus and Robertson, pp. 461–465.

Chase, HB & Silver, AF 1969, 'The biology of hair growth', in Bittar, EE (ed), *The Biological Basis of Medicine*, Volume 6, New York: Academic Press, pp. 3–19.

Cloete, E, Khumalo, NP & Ngoepe, MN 2019, 'The what, why and how of curly hair: a review', https://doi.org/10.1098/rspa.2019.0516.

Cook, R, Evett, I, Jackson, G, Jones, P & Lambert, J 1998, 'A hierarchy of propositions: deciding which level to address in casework', *Science and Justice*, vol. 38, pp. 231–239.

Cotsarelis, G, Sun, T-T & Lavker, RM 1990, 'Label-retaining cells reside in the bulge area of the pilosebaceous unit: implications for follicular stem cells, hair cycle and skin carcinogenesis', *Cell*, vol. 61, pp. 1329–1337.

Crocker, EJ 1991, 'Trace evidence', in Chayko, GM, Gulliver, ED & MacDougall, DV (eds), *Forensic Evidence in Canada*, Aurora, Ontario: Canada Law Book, pp. 259–299.

Cruz, CF, Costa, C, Gomes, AC, Matama, T & Cavaco-Paulo, A 2016, 'Human hair and the impact of cosmetic procedures: a review on cleansing and shape-modulating cosmetics', *Cosmetics*, vol. 3, p. 26. doi:10.3390/cosmetcis3030026.

Cunha, E & Ubelaker, DH 2020, 'Evaluation of ancestry from human skeletal remains: a concise review', *Forensic Sciences Research*, vol. 5, pp. 89–97.

Dachs, J, McNaught, I & Robertson, J 2003, 'The persistence of human scalp hair on clothing fabrics', *Forensic Science International*, vol. 138, pp. 27–36.

Datta, AK, Ghosh, T, Nayak, K & Ghosh, M 2008, 'Menkes kinky hair disease: a case report', *Cases Journal*, vol. 1, p. 158. https://doi.org/10.1186/1757-1626-1-158.

Deedrick, DW & Koch, SL, 2004, 'Microscopy of hair part 1: a practical guide and manual for human hairs', *Forensic Science Communications*, vol. 6, no. 1.

Dror, IE 2011, 'The paradox of human expertise: why experts get it wrong', in Kapur N (ed), *The Paradoxical Brain*, Cambridge: Cambridge University Press, pp. 177–188.

Dror, IE 2016, 'A hierarchy of expert performance', *Journal of Applied Research Memory Cognition*, vol. 5, pp. 121–127.

Dror, IE 2020, 'Cognitive and human factors in expert decision making: six fallacies and eight sources of bias', *Analytical Chemistry*, vol. 92, pp. 7998–8004.

Dror, IE, Thompson, WC, Meissner, CA, Kornfield, I, Krane, D, Saks, M & Risinger, M 2015, 'Context management toolbox: a linear sequential unmasking (LSU) approach for minimizing cognitive bias in forensic decision making', *Journal of Forensic Sciences*, vol. 60, pp. 1111–1112.

Ebling, FJ 1980, 'The physiology of hair growth', in Breuer, MM (ed) *Cosmetic Science*, Volume 2, London: Academic Press, pp. 181–232.

Edson, J, Brooks, EM, McLaren, C, Robertson, J, McNevin, D, Cooper, A & Austin, JJ 2013, 'A quantitative assessment of a reliable screening technique for the STR analysis of telogen hair roots', *Forensic Science International Genetics*, vol. 7, pp. 180–188.

Evans, DJ, Leeder, JD, Rippon, JA & Rivett, DE 1985, 'Separation and analysis of surface lipids of the wool fibre', in *Proceedings of the 7th International Wool Textile Conference, Tokyo*, Volume 1, pp. 181–193.

Fischer, E & Saller, K 1928, 'Eine neue haarfarbentafel', *Anthropologischer Anzliger*, vol. 5, pp. 228–344.

Fraser, IEB 1969, 'Proteins of keratin and their synthesis', *Australian Journal of Biological Sciences*, vol. 22, pp. 231–238.

Fraser, RDB, MacRae, TP & Rogers, GE 1972, *Keratins: Their Composition, Structure and Biosynthesis*, Springfield, Ill: Charles C. Thomas.

Fuentes, A, Ackermann, RR, Athreya, S, Bolnick, D, Lasisi, T, Lee, S, Mc Lean, S & Nelson, R 2019, 'AAPA statement on race and racism', *American Journal of Physical Anthropology*, vol. 169, pp. 400–402.

Gaudette, BD 1976, 'Some further thoughts on probabilities and human hair comparisons', *Journal of Forensic Sciences*, vol. 21, pp. 514–517.

Gaudette, BD 1999, 'Evidential value of hair examination', in Robertson J (ed), *Forensic Examination of Hair*, London: Taylor and Francis, pp. 243–260.

Gaudette, BD & Keeping, ES 1974, 'An attempt at determining probabilities in human scalp hair comparison', *Journal of Forensic Sciences*, vol. 19, pp. 599–606.

Gill, P 2001, 'Application of low copy number DNA profiling', *Croatian Medical Journal*, vol. 42, pp. 229–232.

Gill, P 2014, *Misleading DNA Evidence. Reasons for Miscarriages of Justice'*, London: Elsevier, Academic Press.

Gill, P & Werrett, DJ 1987, 'Exclusion of a man charged with murder by DNA fingerprinting', *Forensic Science International*, vol. 35, pp. 145–148.

Glaister, J 1931, *A Study of Hairs and Wools Belonging to the Mammalian Group of Animals, Including a Special Study of Human Hair, Considered from the Medico-Legal Aspects*, Cairo: MISR Press.

Glucksmann, A 1951, 'Cell deaths in normal vertebrate ontogeny', *Biological Review Cambridge Philanthropic Society*, vol. 26, pp. 59–86.

Guohua, W, Bhushan, B & Torgerson, PM 2005, Nanomechanical characterization of human hair using nanoindentation SEM, *Ultramicroscopy*, vol. 105, pp. 248–266.

Hackett, E 1984, 'Hair', *The Medical Journal of Australia*, vol. 141, p. 300.

Haines, A & Linacre, A 2016, 'A rapid screening method using DNA binding dyes to determine whether hair follicles have sufficient DNA for successful profiling', *Forensic Science International*, vol. 262, pp. 190–195.

Harding, H & Rogers, G 1999, 'Physiology and growth of human hair', in Robertson, J (ed), *Forensic Examination of Hair*, London: Taylor and Francis, pp. 1–77.

Hassall, D, Brealey, N, Wright, W, Hughes, S, West, A, Ravindravath, R, Warren, F & Daley-Yates, P 2018, 'Hair analysis to monitor adherence to prescribed chronic inhaler drug therapy in patients with asthma or COPD', *Pulmonary Pharmacology & Therapeutics*, vol. 51 pp. 59–64.

Hausman, LA 1925, 'The relationships of the microscopic structural characteristics of human head-hair', *American Journal of Physical Anthropology*, vol. 8, pp. 173–177.

Hellmann, A, Rohleder, U, Schmitter, H & Wittig, M 2001, 'STR typing of human telogen hairs – a new approach', *International Journal of Legal Medicine*, vol. 114, pp. 269–273.

Herrington, V & Colvin, A 2015, 'Police leadership for complex times', *Policing: A Journal of Policy and Practice*, vol. 10, pp. 7–16.

Hicks, JW 1977, *Microscopy of Hairs: A Practical Guide and Manual*, Washington D.C.: Federal Bureau of Investigation, FBI Laboratory.

Hietpas, J, Buscaglia, J, Richard, AH, Shaw, S & Castillo, HS 2016, 'Microscopical characterization of known postmortem root bands using light and scanning electron microscopy', *Forensic Science International*, vol. 267, pp. 7–15.

Hofbauer, GFL, Tsambaos, D, Spycher, MA & Tröeb, RM 2001, 'Acquired hair fragility in Pili anulati: causal relationship with androgenetic alopecia', *Dermatology*, vol. 203, pp. 60–62.

Houck, MM & Budowle, B 2002, 'Correlation of microscopic and mitochondrial DNA hair comparisons', *Journal of Forensic Sciences*, vol. 47, pp. 964–967.

Howes, LM, Kirkbride, KP, Kelty, SF, Julian, R & Kemp, N 2013, 'Forensic scientists conclusions: how readable are they for nonscientists report-users?', *Forensic Science International*, vol. 231, pp. 102–112.

Ito, S & Wakamatsu, K, 2011, 'Diversity of human hair pigmentation as studied by chemical analysis of eumelanin and pheomelanin', *Journal of the European Academy of Dermatology and Venereology*, doi: org:10.1111/j.1468-3083.2011.04278.

Jablonski, NG 2010, 'The naked truth', *Scientific American*, February, pp. 28–35.

Jackson, G, Aitken, C & Roberts, P 2014, 'Case assessment and interpretation of expert evidence', *Practitioner Guide No. 4, Working Group on Statistics and Law of the Royal Statistics Society*.

James, WD, Elston, D & Berger, T 2006, *Andrews' Diseases of the Skin Clinical Dermatology Tenth Edition*. Saunders/Elsevier, p. 299. https://doi.org/10.1016/j.ultramic.2005.06.033.

Jimbow, K, Fitzpatrick, TB & Wick, MM 1991, 'Biochemistry and physiology of melanin pigmentation', in Goldsmith, LA (ed), *Physiology, Biochemistry and Molecular Biology of Skin*, Oxford: Oxford University Press, pp. 873–909.

Kelty, S, Julian, R & Robertson, J 2009, 'Professionalism in crime scene examination: the seven key attributes of top crime scene examiners' *Forensic Science Policy Manual*, vol. 2, pp. 175–186.

Kind, SS & Owens, GW 1977, 'Assessment of information content gained from the microscopical comparison of hair samples', *Journal of Forensic Sciences Society*, vol. 16, pp. 235–239.

King, LA, Wigmore, R & Twibell, JM 1982, 'The morphology and occurrence of human hair sheath cells', *Journal of the Forensic Science Society*, vol. 22, pp. 267–269.

Kligman, A 1962, 'Facts and fancies on the care of the hair and nails', *Southern Medical Journal*, vol. 55, pp. 1011–1020.

Koch, SL, Liebowitz, BS & Shriver, MD 2020, 'Microscopical discrimination of human head hairs sharing a mitochondrial haplogroup', *Journal of Forensic Sciences*, https://doi.org/10.1111/1556-4029.14560.

Lambert, M & Balthazard, V 1910, *Le Poil de l'homme et des animaux. Applications aux expertises medico-legales et aux expertises des fourrures*, Paris: Steinheil.

Lawton, ME & Sutton, JG 1982, 'Multiple enzyme typing of sheath cells associated with the root of a single human head hair', *Journal of the Forensic Science Society*, vol. 22, pp. 203–209.

Lee, LD, Ludwig, K & Baden, HP 1978, 'Matrix proteins of human hair as a tool to identify individuals', *Forensic Science*, vol. 11, pp. 115–121.

Lee, SY, Ha, EJ, Woo, SK, Lee, SM, Lim, KH & Eom, YB 2017, 'A rapid nuclear staining test using cationic dyes contributes to efficient STR analysis of telogen hair roots', *Electrophoresis*, https://doi.10.1002/elps.201700024.

Linch, CA & Prahlow, JA 2001 'Postmortem microscopic changes observed at the human head hair proximal end', *Journal of Forensic Sciences*, vol. 46, pp. 15–20.

Loussouarn, G 2001, 'African hair growth parameters', *British Journal of Dermatology*, vol. 145, pp. 294–297.

Lucas, DM 2007, 'Report on forensic science matters to the commission of inquiry Re: James Driskell', *Appendix G in 'Report of The Commission of Inquiry into Certain Aspects of the Trial and Conviction of James Driskell*.

Marquis, R, Biedermann, A, Cadola, L, Champod, C, Guueissaz, L, Massonnet, G, Mazella, WD, Taroni, F & Hicks, T 2016, 'Discussion on how to implement a verbal scale in a forensic laboratory: benefits, pitfalls and suggestions to avoid misunderstandings', *Science and Justice*, vol. 56, pp. 364–370.

Marshall, RC, Gillespie, JM & Klement, V 1985, 'Methods and future prospects for forensic identification of hairs by electrophoresis', *Journal of Forensic Sciences Society*, vol. 25, pp. 57–66.

Martire, KA & Watkins, I 2015, 'Perception problems of the verbal scale; A reanalysis and application of a membership function approach', *Science and Justice*, vol. 55, pp. 264–273.

Martire, KA, Kemp, RI, Sayle, M & Newall, BR 2014, 'On the inter-
    pretation of likelihood ratios in forensic science evidence: pre-
    sentation formats and the weak evidence effect', *Forensic Science
    International*, vol. 240, pp. 61–68.

Marx, V, 2013, 'Biology: the big challenges of big data', *Nature*, vol. 498,
    pp. 255–260.

McNevin, D, Wilson-Wilde, L, Robertson, J, Kyd, J & Lennard, C 2005,
    'Short tandem repeat (STR) genotyping of keratinized hair. Part 2.
    An optimized genomic DNA extraction procedure reveals donor
    dependence of STR profiles', *Forensic Science International*, vol. 153,
    pp. 247–259.

McNevin, D, Edson, J, Robertson, J & Austin, JJ 2015, 'Reduced reac-
    tion volumes and increased Taq DNA polymerase concentration
    improve STR profiling outcomes from a real-world low template
    DNA source: telogen hairs', *Forensic Science Medical Pathology*,
    vol. 11, pp. 326–338.

Melton, T, Dimick, G, Higgins, B, Lindstrom, L & Nelson, K 2005,
    'Forensic mitochondrial DNA analysis of 691 casework hairs',
    Journal of Forensic Sciences, vol. 50, pp. 73–80.

Melton, T, Dimick, G, Higgins, B, Yon, M & Holland, C 2012a,
    'Mitochondrial DNA analysis of 114 hairs measuring less than 1 cm
    from a 19-year-old homicide', *Investigative Genetics*, vol. 3, p. 12.
    https://www.investigativegenetics.com/content/3/1/12.

Melton, T, Holland, C & Holland, M 2012b, 'Forensic mitochondrial
    DNA analysis: current practice and future potential', *Forensic
    Science Review*, vol. 24, pp. 101–122.

Menkes, JH, Alter, M, Steigleder, GK, Weakley DR & Ho Sung, J 1962,
    'A sex-linked recessive disorder with retardation of growth, pecu-
    liar hair, and focal cerebral and cerebellar degeneration', *Pediatrics*,
    vol. 29, pp. 764–779.

Montagna, W 1963, 'Phylogenetic significance of the skin of man',
    Archives of Dermatology, vol. 88, pp. 1–19.

Montagna, W 1976, 'General review of the anatomy, growth, and develop-
    ment of hair in man', *in Biology and Disease of the Hair*, Baltimore:
    University Park Press, pp. xxi–xxxi.

Montagna, W & Van Scott, EJ 1958, 'The anatomy of the hair follicle',
    in Montagna W & Ellis, RA, *Biology of Hair Growth*, New York:
    Academic Press, pp. 1–32.

Montagna, W & Parakkal, P 1974, 'The pilary apparatus', *The Structure and
    Function of Skin*', 3rd edn, New York and London: Academic Press.

Mullen, C, Spence, D, Moxey, L & Jamieson, A 2014, 'Perception prob-
    lems of the verbal scale', *Science and Justice*, vol. 54, pp. 154–158.

Murch, RS & Budowle, B 1986, 'Applications of isoelectric focusing in
    forensic serology', *Journal of Forensic Sciences*, vol. 31, pp. 869–880.

Myers, RJ & Hamilton, JB 1951, 'Regeneration and rate of growth of hairs in man', *Annals of the New York Academy of Sciences*, vol. 53, pp. 562–568.

Netherton, EW 1958, 'A unique case of trichorrhexis nodosa: 'bamboo hairs'', *AMA Archives of Dermatology*, vol. 78, pp. 483–487.

Oliver, RF & Jahoda, CAB 1981, 'Intrafollicular interactions', in Orfanos, EC, Montagna, W & Stuttgen, G (eds), *Hair Research*, Berlin & Heidelberg: Springer-Verlag, pp. 18–24.

Orentreich, N 1969, 'Scalp hair replacement in man', in Montagna, W & Dobson, RL, (eds), *Advances in Biology of Skin. Hair Growth*, Volume 9, Oxford: Pergamon, pp. 99–108.

Ottens, R, Taylor, D, Abarno, D & Linacre, A 2013, 'Successful direct amplification of nuclear markers from a single hair follicle', *Forensic Science Medicine and Pathology*, vol. 2, pp. 238–243.

Parker, GJ, Leppert, T, Anex, DS, Hilmer, JK, Matsunami, N, Baird, L, Stevens, J, Parsawar, K, DurbinJohnson, BP & Rocke, DM 2016, 'Demonstration of protein-based human identification using the hair shaft proteome', *PLOS ONE*, DOI:10.1371/journal.pone.0160653.

Pangerl, E & Igowsky, K 2007, *Changes Observed in Human Head Hairs Exposed to Heat*, Minnesota Bureau of Criminal Apprehension, pp. 1–14. Pangerl.doc (nfstc.org).

Pelfini, C, Cerimele, D & Pisnau, G 1969, 'Aging of the skin and hair growth in man', in Montagna, W & Dobson, RL, (eds), *Advances in Biology of Skin. Hair Growth*, Volume 9, Oxford: Pergamon, pp. 211–220.

Piccolo, V, Cirocco, A, Russo, T, Piraccini, BM, Starace, M, Ronchi, A & Argenziano, G 2018, 'Hair cross-sectioning in uncombable hair syndrome: an easy tool for complex diagnosis', *Journal of the American Academy of Dermatology*, vol. 79, pp. E63–E64.

Pinkus, H, 1958, 'Embryology of hair', in Montagna, W & Ellis, RA, (eds), *The Biology of Hair Growth*, New York: Academic Press, pp. 1–32.

Prokopec, M, Glasova, L & Ubelaker, DH 2000, 'Change in hair pigmentation in children from birth to 5 years in a central European population (longitudinal study)', *Forensic Science Communications*, vol. 2, pp. 1–2.

Righetti, PG 2005, 'Electrophoresis: the march of pennies, the march of dimes', *Journal of Chromatography Archives*, vol. 1079, pp. 24–40.

Robertson, J 1982, 'An appraisal of the use of microscopic data in the examination of human head hair', *Journal of Forensic Sciences Society*, vol. 22, pp. 390–395.

Robertson, J 1999a, (ed), *Forensic Examination of Hair*, London: Taylor and Francis.

Robertson, J 1999b, 'Forensic and microscopic examination of human hair', in Robertson, J (ed), *'Forensic Examination of Hair'*, London: Taylor and Francis, pp. 79–154.

Robertson, J & Aitken, CGG, 1986, 'The value of microscopic features in the examination of human head hairs: analysis of comments contained in questionnaire returns', *Journal of the Forensic Sciences*, vol. 31, pp. 563–573.

Robbins, CR 1988, *Chemical and Physical Behavior of Human Hair*, 2nd edn, New York: Springer-Verlag.

Roe, GM, McArdle, W & Pole, K 1985, 'Detection of cosmetic treatments on human scalp hair. Screening of forensic casework samples', In: *Proceedings of the International Symposium of Forensic Hair Comparisons*', Washington, DC: Government Printing Office, pp. 63–68.

Roe, GM, Cook, R & North, C 1991, 'An evaluation of mountants for use in forensic hair examination, *Journal of the Forensic Science Society*, vol. 31, pp. 59–65.

Rogers, GE 1964, 'Structural and biochemical features of the hair follicle', in Montagna, W & Lobitz, W C (eds), *The Epidermis*, New York: Academic Press, pp. 179–236.

Rook, AR & Dawber, R 1982, *Diseases of the Hair and Scalp*, Oxford: Blackwell Scientific.

Roux, C, Talbot-Wright, B, Robertson, J, Crispino, F & Ribaux, O, 2016, 'The end of the (forensic science) world as we know it? The example of trace evidence', *Philosophical Transaction of the Royal Society B*, vol. 370, 20140260, DOI:10.1098/rstb.2014.0260.

Rowen, TS, Gaither, TW, Awad, MA, Osterberg, E, Charles, SAW & Breyer, BN, 2016, 'Pubic hair grooming prevalence and motivation among women in the United States ', *Journal of the American Medical Association Dermatology*, vol. 152, 1106. doi:10.1001/jamadermatol.2016.2154.

Rudnicka, L, Rakowska, A, Olszewska, M, Slowinska, M, Czuwara, J, Rusek, M & Costa Pinheiro, AM 2012, *Atlas of Trichoscopy*, London: Springer. DOI: https://doi.org/10.1007/978-1-4471-4486-1

Saitoh, M, Uzuka, M, Sakamoto, M & Kobori, T 1969, 'Rate of hair growth', in Montagna, W & Dobson, RL, (eds), *Advances in Biology of Skin. Hair Growth*, Volume 9, Oxford: Pergamon, pp. 183–201.

Seta, S, Sato, H & Miyake, B 1988, 'Forensic hair investigation', in Maehly, A & Williams RL (eds), *Forensic Science Progress*, Volume 2, New York: Springer-Verlag Berlin.

Sims, RJ 1967, 'Hair growth in Kwashiorkor', *Archives of Diseases of Childhood*, vol. 42, pp. 397–400.

Smith, S 1982, '*Mostly Murder. An Autobiography*', London: Harrap.

Spector, DL & Goldman, RD 2008 (eds), *Basic Methods in Microscopy*, Cold Spring Harbor Laboratory Press, Cold Spring Harbor Protocols.

Straile, WE 1965, 'Root sheath-dermal papilla relationships and the control of hair growth', in Lyne, AG & Short, BF (eds), *Biology of the Skin and Hair Growth*, Sydney: Angus and Robertson, pp. 35–57.

Sturm, R 2009, 'Molecular genetics of human pigmentation diversity', *Human Molecular Genetics*, vol. 18, pp. R9–R17.

Suter, D 1979, 'Hair colour in the Faroe and Orkney Islands', *Annals of Human Biology*, vol. 6, pp. 89–93.

Swift, JA 1977, 'The histology of keratin fibers', in Asquith, RS (ed), *Chemistry of Natural Protein Fibers*, New York: Plenum Press, pp. 81–146.

Swift, JA 1981, 'The hair surface', in Orfanos, EC, Montagna, W & Stuttgen, G (eds), *Hair Research*, Berlin & Heidleberg: Springer-Verlag, pp. 18–24.

Tafaro, JT 2000, 'The use of microscopic postmortem changes in anagen hair roots to associate questioned hairs with known hairs and reconstruct events in two murder cases', *Journal of Forensic Sciences*, vol. 45, pp. 495–499.

Teerink, BJ 2003, *Hair of West-European Mammals. Atlas and Identification Keys*, Cambridge: Cambridge University Press.

Thibaut, S, Gaillard, O, Bouhanna, P, Cannell, DW & Bernard, BA, 2005, 'Human hair shape is programmed from the bulb', *British Journal of Dermatology*, vol. 152, pp. 32–638.

Thomas, E 1969, 'Search behaviour', *Radiologic Clinic North America*, vol. 7, pp. 403–417.

Tiselius, A 1937, 'A new apparatus for electrophoretic analysis of colloidal mixtures', *Transactions of the Faraday Society*, vol. 33, pp. 524–531.

Tobin, DJ, Hordinsky, M & Bernard, BA 2005, 'Hair pigmentation: a research update', *Journal of Investigative Dermatology Symposium Proceedings*, vol. 10, pp. 275–279.

Tobin, DJ 2009, 'Aging of the hair follicle pigmentation system', *International Journal of Trichology*, vol. 1, pp. 83–93.

Tuddenham, W 1962, 'Visual search, image organization, and reader error in roentgen diseases: studies on the psychophysiology of roentgen image perception', *Radiology*, vol. 78, pp. 694–704.

Vesterberg, O 1989, 'History of electrophoretic methods', *Journal of Chromatography Archives*, vol. 480, pp. 3–19.

Vincent, F 2010, 'Report: inquiry into the circumstances that led to the conviction of Mr. Farah Abdulkadir Jama', http://www.parliament.vic.gov.au/papers/govpub/VPARL2006-10No301. pdf 35.

von Beroldingen, CH, Roby, RK, Sensabaugh, GF & Walsh, S 1989, 'DNA in hair', *Proceedings of the International Symposium on the Forensic Aspects of DNA analysis, Quantico, 1989.* Washington, DC: US Government Printing Office, pp. 265–266.

Wickenheiser, RA & Hepworth, DG 1990, 'Further evaluation of probabilities in human scalp hair comparisons', *Journal of Forensic Sciences*, vol. 35, pp. 1323–1329.

Wortmann, FJ, Wortmann, G & Sripho, T, 2020, 'Why is hair curly? – Deductions form the structure and biomechanics of the mature hair shaft', *Experimental Dermatology*, vol. 29, pp. 366–372.

Wyllie, AH 1980, 'Glucocorticoid-induced thymocyte apoptosis is associated with endogenous endonuclease activation', *Nature*, vol. 284, pp. 555–556.

Zahn, H, Messinger, H & Hocker, H 1994, 'Covalently-linked fatty acids at the surface of wool: part of the cuticle cell envelope', *Journal of Textile Research*, vol. 64, pp. 554–555.

Zviak, C & Dawber, RPR 1986, 'Hair structure, function and physiochemical properties', in Zviak, C (ed), *The Science of Hair Care*, New York and Basel: Marcel Dekker.

Referenced Standards

Australian Standards (AS)

AS

5388 Forensic Analysis

5388.1 Part 1: Recognition, recording, recovery, transport and storage of material

5388.2 Part 2: Analysis and examination of material

5388.3 Part 3: Interpretation

5388.4 Part 4: Reporting

5483 Minimizing the risk of contamination in products used to collect and analyze biological material for forensic DNA purposes

International Standards

International Organisation for Standardization (ISO)/International Electrotechnical Commission (IEC)

17025:2017 General requirements for the competence of testing and calibration laboratories

3100 Risk management—Principles and guidelines

21043 Forensic Sciences

21043-1 Part 1: Terms and definitions

21043-2 Part 2: Recognition, recording, collecting, scene control, transport and storage of items

(Note, Part 3: Analysis, Part 4: Interpretation and Part 5: Reporting are under development as of December 2020)

ISO 27037 Information technology—Security techniques—Guidelines for the identification, collection, acquisition and preservation of digital evidence

Index

Note: Page numbers in *italics* indicates figures and **bold** indicates tables in the text.

A

Abraded tip, *40*, *82*, *86*; *see also* Tip
Acquired characteristics, 10, 13, 72, 110, 189–208
  arson/explosion, 201–203
  arson/fire *versus* cosmetic heat styling, 200–201
  blunt-bladed implement, 195–196
  blunt force impact, 198–200
  broken/crushed hair, 198–199
  cut hairs, 192–193
  damage checklist, **191**
  examinations with, 190–191
  explosion, 203–205
  floor safe door, 199–200
  head lice, 205–207, *207*
  motor vehicle collision, 193–195
  natural and unnatural hair loss, 192
  overview, 189–190
  reporting of, 191–192
  sharp-bladed implement, 196–198
  windscreen impact, 193–195, *196*
Adenosine triphosphate (ATP), 30
Aeriform lattice, *68*, 119; *see also* Lattice
African, 71
  hair colour, 46
  hair shaft, **72**
  less dense hair coverage than Caucasian, 26, 37
  scalp hairs, 39, 69, 71, 77, 179
Aitken, C. G. G., 7, 14, 93
A layer, 41
Alternative hypothesis, 129, 134, *134*, 153; *see also* Null hypothesis

American Association of Physical Anthropologists, 71
American Journal of Physical Anthropology, 5
American Naturalist, 6
Anagen, 7, 10–11, 29–30
  follicle, 34, *34*
  hairs, 28, 30, 32–33, 54, 74–75, 80–82, 86, 204, *205*
  length, 33
  representation of hair follicles, *33*
  roots, *35*, *82*, *84*, 191–192, *205*, 208
  scalp hairs in, 36
Anatomy
  cortex, 43–45
  cuticle, 41–43, *42*
  medulla, 45
Animal hairs
  colour banding, *65*
  compared with human hairs, **66**
  examination proforma, 114–115
  features, 119
  medulla, 31, *68*, 104
  non-human, 25, 45
  recognition of, 63–69
  scale pattern, 110
  separation of, 63–69, 173–177
Apoptosis in hair, 121
Arrector pili muscle, 27, 29, 30, *33*
Arson/explosion, 201–203
Arson/fire, 200–201
Asexual hairs, 37
Asian, 39, 69, 71, **72**, 138, 179
ATP, *see* Adenosine triphosphate
Australian Standard
  AS 5388.3, 130–132
  AS 5388.2-2012, 61–62
  AS 53388.3-2013, 128

AS 5388.4-2013, 139, 141, 144, 146–148, 172
AS5388.1, 49–50, 55
Automaticity, 21
Auxiliary/underarm hairs, **70**
Average probabilities, 6
Average value of hairs, 151

B
Back combing, *58*
Balthazard, V., 4, *5*
Bamboo hair, 88; *see also* *Trichorrhexis invaginate*
Banding
    colour, 65, *65*
    light and dark, 91
    post-mortem, 75, *86*
    root, 81
Barnett, P., 6
Bayesian approach, 7, 152
Bayesian inference, 148, 151
Beard/moustache
    hair, 38, 46, **70**, 73, 109, 178
Bias
    cognitive, 3, 148, 152–153
    confirmation, 16
    context, 87, 130
    potential, 17
    risk of, 130
    sources of, 153
    unconscious, 92
Biasability, 16
Big data, 11; *see also* Data
Bioinformatics, 11
Biological material, 4, 9–11, 13, 23, 55
Biology and chemistry of hairs, 174–175, 181–182
Biometrics, 13
Birbeck, M. S. C., 31
Black/opaque hair, 14, 47, 77, *84*
Blond hair, *see* Yellow hair
Blunt-bladed implement, 195–196
Blunt force impact, 198–200
Boccalatte, S., 26
Body area determination, 39, 69–70, **70**, 73–74, 77, 135, 177–178

Body hair, 37–38, 46, 56–57, 70, 72, 77, 178, 192
Boonen, T., 120
Bourguignon, L., 120
Brandhagen, M. D., 74, 76
Broken (BR)
    hair, 198–199
    tip, 44, 82, *86*
Brooks, E. M., 15, 75–76, 192
Brough, I., 192
Brown hair, *35*, *81*, *85*, *96–99*, *113*, 147, 200
Brunner, H., 67
Brush ends, 42
Budowle, B., 7, 12, 17, 74
Bulbous hair peg, 27, *28*
Bulge, 27, 30, 33, *33*, 88

C
Canities, 46
Case management, 12–17, 23, 62, 172
Case study, 73–75
Catagen, 10, 27, 29–30, 32, *33–35*, 34, 36, 74–75, 80, 180
Caucasian, 26, 37, 39, 46, 63, 71, **72**, 147
Cell biologists, 25
Cell differentiation, 27, 31, 95
Cell division, 25, 30, 34, *35*
Cell membrane complex (CMC), 32, 43
Cell proliferation, *29*, 31
Certificates, 141
Chao, J., 126
Chase, H. B., 38
Checklist, 14–15
    for assessing colour, 46
    damage, **191**
    developing, 93
    reveal damage trends, 191
    TLM level of examination, 110
    for visual and/or LPM examinations, 76
Chemical fixation, 121
Chest hairs, *40*, **70**, 198
Cognitive bias, 3, 148, 152–153; *see also* Bias

Colour, 77–80
  abnormal, 19
  acquired, 14
  animal hairs, 119
  artificial, 78–79, *85*, *108*,
      111–112, 126
  assessment, 46–47, 77, 95
  banding, 65, *65*
  banding in animal hair, *65*
  black, 14, 47, 77
  brown, 14, 47, 77, *85*, *95*, 147,
      160, 167
  changing, 26
  classifying, 15
  colourless hairs, 16, *79*
  cortical cells, 45
  female arm and leg hairs, *40*
  greyish, *82*
  perceived, 46
  and pigmentation, 46–47, **66**, 80
  red, 14, 47, 77, *83*, *97*, *99*, 199
  rinse, *79*
  unmounted hairs, 77
  visual hair, 14
  white, *79*
  yellow, 14, 47, 77, *80*, *85*
Colvin, A., 22
Combing, 34, 54, 57, 58, 110, 190
Committee on air Examination, 19
Committee on Forensic Hair
      Comparison (CFHC), 93
Comparison microscopy, 6, 110–113,
      121, 135–136, 180–188;
      see also Microscopy
Confirmation bias, 16; *see also* Bias
Confirmatory test, 17, 136
Contemporary knowledge, 131
Context bias, 87, 130
Contextual information, 130–131
Cook, R., 10
Cortex, 30, 39, 43–45, 65
  cortical cells, *199*
  cut-away section, *41*
  and cuticle, 15, 29, 31, *33*, *206*
  distinctive characteristics, *102*
  medulla and, 93
Cortical fusi, 15, 35, 45, 79, *82*,
      95–97, *96*, *108*, 137, 147,
      204

Cortical texture, 44, *80*, *83*, 97–104,
      99, *108*, *113*, 147, 159
Cosmetic heat styling, 200–201
Counting stained nuclei, 124–125
Cracked cuticle, 206
Crime scene examiner (CSE), 12,
      49–51, 54–57
Criminalistics, 9, 12–17, 23, 27, 57,
      138, 143, 183
Crocker, E. J., 145–146
Cross examination, 156–157
Cross-section, 89, 119
Crushed and broken hair
      shafts, *199*
Crushed tip, 82
Cruz, C. F., 69
Cunha, E., 71
Curly hair, 77, 89
Current competence, 137
Cut hairs, 54, *58*, 79, 192–193,
      196–197
Cuticle, 30, 39, 41–43, *42*, *107*
  anatomy, 41–43, *42*
  cracked, 206
  germinative tissues, 32
  layer, *103*, *111*
  looped, 110
  pigment distribution, *98*
  ragged, 110
  scale pattern of human hair, *42*
  and scales, 110
  serrated, 110
  smooth, 110
  thicknesses, *111*
  types, *111–112*
Cut tip, 192, *201*

D
DAPI-DABCO staining, 124
DAPI (40-6 diamidino-
      2-phenylindole), 120
  stained telogen roots, *122–123*
Data
  big, 11
  capture and analysis, 94
  deconvolute, 11
  expertise and, 16
  frequency, 132
  mt-DNA, 74

qualitative/quantitative, 130, 144
transformation, 127–131
Daubert, 22
Dawber, R., 46, 83
Decision-making, 13, 61, 74
Deedrick, D. W., 69
Degrees of certainty, 146
Dendrites, 30–31, 47
Dermal papilla, 27, 30–32, *33–34*,
45, 47
Directional friction, 42
Discontinuity, 45
DNA analysis, 9–10, 12, 17, 23,
69–84, 94, 121, 125, 135,
138, 144, 189
Driskell Inquiry, 1–2, 14, 19, 128
Dror, I. E., 16–17, 21–22, 152–153
Dyed hair, 79, *85*, 200

E
Early outcomes, 13
Ebling, F. J., 27
Effective outcomes, 13
Electron micrographs of tool
signature cuts, *194*
Electrophoresis, 7
E-mails, 139–140
Endocuticle, 41
Epicuticle, 42; *see also* Cuticle
Era 1 of hair examination, 4–6
Era 2 of hair examination, 6–9
Era 3 of hair examination, 9–11
Era 4 of hair examination, 11–12
Era of observation, 4
Error, 133–138
examiner, 6
human, 3
statistical, 21, 133
Error type 1, 12, 18, 23, 74, 133–136,
134, 180, 183–184, 186
Error type 2, 12, 18, 23, 133–134,
134, 136, 180, 183–184, 186
Error type 3, 18
Ethnic origin
determinations, 178–179
of human hairs, 71–72, **72**
Eumelanin, 14, 47
European, 71, 178

European Network of Forensic
Science Institutes (ENFSI),
148, 171
Evaluation, 127–138
error, 133–138
estimating probabilities, 132–133
formulating an opinion, 132
professional judgement, 148, *149*
reporting, 148
stage, 148, *149*
transforming data into
information, 127–131
Evidence recognition, 49
Examinations, 1–23, 145–147,
179–190; *see also*
Laboratory examinations
with acquired characteristics,
190–191
case management, 12–17
criminalistics value, 12–17
cross examination, 156
detailed examination, 84–113
comparison microscopy,
110–113
high-power transmitted light
microscopic examination,
92–110
discriminating power of, 8
FBI approach to hair reporting,
18–22
forensic hair, 188
history, 4–12
of human hairs, 69–84
body area determination,
69–70, **70**
case study, 73–75
ethnic origin, 71–72, **72**
low-power microscopic (LPM),
76–83, *84*
selection for DNA analysis,
72–76
implications, 18–22
low-power microscopic examination
(LPM), 172–173,
175–176, 177, 182–183
protocol for hair, 157–158,
161, 165, 168–169
report/reporting, 157–169

Exocuticle, 41
Expert witness, 153–157
Explosion, 203–205
Extraneous physical materials, 54;
        see also Physical material
Eyewitness identification, 73

F
Fatty acid, 30
Federal Bureau of Investigation (FBI)
    approach to hair reporting, 18–22
    meeting hosted by, 8
    testimony on microscopic hair
        analysis, 4
Fischer-Saller Scale, 14, 46
Fit for purpose, 62, 127, 134–136,
        140
Fixation, 121, 124
Follicles, 25–26
    anagen, 34
    density, 36–37
    development, 33
    formation, 27–29
    lower, 30
    replacing, 26
    role in growth of hair fibres,
        29–32
    telogen, 34
    upper, 30
Follicular tag, 34
Forensic examiner, 6, 12, 14, 83
Forensic hair examination, 7, 22,
        126, 188–191
Forensic medical examination kit
        (FMEK), 55
Forensic medical officer (FMO), 55–56
Forensic medicine, 6
Forensic nursing, 56
Forensic reference hair collection kit,
        57, 59
Formal statement, 142
Formulating opinion, 132
Foundational validity, 3–4, 9, 12
Fourier transform infrared
        spectroscopy (FTIR), 171
Frayed tip, 82
Frequency data, 132
Full report, 140–143

G
Gaudette, B., 6–8, 19–20, 128
Germinative cells, 27, 30, 32, 45
Get It First Time (GIFT principle), 54
Gill, P., 9
Glaister, J., 6
Greyish hair, 82
Growth cycle, 25, 27, 32, 33–34, 35,
        79, 160, 175, 180, 182
Guard-hairs, 63, 64

H
Haematoxylin, 75, 120–121, 123,
        123–125, 125
Hair
    abnormal cuts in, 197
    anagen roots, 205
    anatomy, 39–45
        cortex, 43–45
        cuticle, 41–43, 42
        medulla, 45
    arson/fire versus cosmetic heat
        styling, 200–201
    average value of, 151
    biology and chemistry of,
        174–175, 181–182
    blunt-bladed implement, 195–196
    blunt force impact, 198–200
    classification, 46
    colour, see Colour
    comparison microscopy, 180–188
    cone, 27
    crushed and broken shafts, 199
    cutting, 192–193, 193
    cut with electric razor, 202
    cycle, 28–30, 32–36
    detailed examination of, 180–188
    distribution, 36–37
    electron micrographs of tool
        signature cuts, 194
    examination, see Examinations
    follicle, see Follicle
    forensic examination, 188
    germ, 27, 28
    growth cycle, 34
    growth rates, 37–38
    length, 76–77
    loss, 25, 192

microscopic features of at the LPM
    level, 182–183
microscopic features of at the
    TLM level, 183–185
morphology, 38–39
motor vehicle collision, 193–195
peg, 27, *28*
pigmentation, 46–47, 95
principles of microscopy and
    sample preparation, 175–176
recovery and collection of,
    173–174
report, 18–22, 187–188
selection for DNA testing,
    177–180
shaft abnormalities, **87**
shaft cut with kitchen knife, *198*
shaft profiles, *78*
sharp-bladed implement, 196–198
statistics and verbal scales for,
    148–152
structure of human and animal,
    176–177
tip, *44, 86, 203*
training, 172
types, 37, **66**
windscreen impact, 193–195, *196*
"Hair" (Boccalatte and Jones), 26
Hair comparison
    forensic, 6
    microscopic, 2, 18
    negative, 152
    positive, 151–152
Hairdressing, 192, *193*
"Hair ID, " 67
Hairless, 26
Hamilton, J. B., 38
Handpicking hairs, 51
Hardening, 31
Harding, H., 25–27, 31–32, 37,
    42–43, 45–47
Harris's haematoxylin, 120–121, 124
Hassall, D., 46
Hausman, L. A., 5
Head lice, 205–207, *207*
Henle layer, 32
Hepworth, D. G., 6
Herrington, V., 22

Hicks, J. W., 38, 69, 71
Hierarchy of propositions, 145
Hietpas, J., 81
High-power transmitted light
    microscopic examination,
    92–110
Hispanic, 71
Histomount™, 14
Houck, M. M., 12, 17, 74
Howes, L. M., 146
Human hairs; *see also* Hair
    compared with animal hairs, **66**
    examination proforma
        comparison, 118
        LPM, 116
        TLM, 117
    general features of, **70**
    recognition of, 63–69, 173–177
    root growth phases of, *35*
    seperation of, 63–69, 173–177
Huxley layer, 32
Hydrophobicity, 42
Hypothesis, *see* Alternative hypothesi
Hystomount™, 66

I
Igowsky, K., 200, 203, 204
Imbricate cuticle scale pattern of
    human hair, *42*
Inconclusive conclusion, 144
Individualisation, 2, 12, 14, 138
Inherent characteristics, 10
Inner root sheath (IRS), 27, 31–32
Interim report, 140; *see also* Report/
    reporting
International Association of Forensic
    Sciences (IAFS), 8
Interpretation, *see* Evaluation
Intimate sample, 55
Ironic rebound, 153
Isoelectric focusing (IEF), 7
ISO/IEC 17025:2005, 61
ISO/IEC 17025:2018, 61, 67

J
Jablonski, N. G., 26
Jones, M. R., 26
Jurisdiction

for CSE, 54
legal requirements, 49, 140,
    154–155

K
Keratin-associated proteins
    (KAP), 43
Keratin gene expression, 31
Keratin intermediate filaments
    (keratin IF), 31, 43
Keratin microfibrils, 43
Keratogenous, 31, 82
Kind, S., 6
Kinky hair, *89, 92*
Knotted hair, *see* Trichonodosis
Known hair, 157–158, 161, 165,
    168–169
Known samples, 14, 16–17, 20,
    54, 62, 77, 87, 113, 134,
    136–138, 144, 152, 193
Koch, S. L., 69, 71, 74, 128

L
Laboratory examinations, 61–126;
    see also Examinations
    level 1, 61, 63–69
    level 2, 61, 69–84
    level 3, 61, 84–113
    level 4, 61
Laboratory report, 18, 140–141
Ladder medulla, 67, 119
Lambert, M., 4, *5*
Lanugo hairs, 27–28, 37
Lattice
    wide medulla, 68
Lee, S. Y., 76
Leica stereo microscope, 63
Level 1 examinations, 134–135
Level 2 examinations, 135
Level 3 examinations, 135–136
Likelihood ratio, 132, 148
Limb hairs, 70
Linch, C. A., 81
Linear sequential unmasking (LSU),
    17, 153
Logical thinking, 148
Looped cuticle, 110; *see also*
    Cuticle

Lower follicle, 30
Low-power microscopic examination
    (LPM), 10, 14–16, 32, 61,
    64, 67, *76–83, 84,* 172–173,
    175–177, 182–183, 193–194
Lucas, D. M., 2

M
Macroscopic animal hair profiles, *64*
Macroscopic assessment of natural
    hair colour, 46–47
Marquis, R., 150
Martire, K. A., 150
Massively parallel sequencing, 11
McNevin, D., 11
Medulla, 30, 39, 65, 107, 119
    anatomy, 45
    animal hairs, 31, *68,* 104
    cortex and, 93
    distributions, *109*
    ladder, 67, 119
    opaque, 104
    translucent, 104
    types, *109*
Medullary index (MI), 104
Melanocytes, 15, 30–31, 47, *79*
Melanosomes, 15, 47, *95*
Melton, T., 11
Mercer, E. H., 31
18-methyl-eicosanoic acid (18-MEA),
    42
Methylene blue staining, 126
Microscopy, 171, 172–173
    comparison, 6, 110–113, 121,
        135–136, 180–188
    era of observation and, 4
    hair analysis, 4
    hair comparison, 3
Microspectrophotometry (MPS), 171
Mitochondrial DNA (mt-DNA), 74
Molecular biologists, 25
Mongoloid, 71
Monoilethrix, 88
Montagna, W., 25, 36
Motor vehicle collision, 193–195
Mt-DNA, 2, 4, 11–12, 17–18, 45,
    74–75, 136, 138, 141,
    180–181

Mullen, C., 150
Multiserial ladder, *68*
Multiseriate ladder, 119
Murch, R. S., 7
Myers, R. J., 38

N
"Naked ape" 26
National Academy of Science (NAS), 22
National Institute of Forensic Sciences (NIFS), 148, 150–151
National Institute of Standards and Technology (NIST), 171
Natural and unnatural hair loss, 192
Natural hair colour, 46–47; *see also* Colour
Natural tip, 82
Negroid, 71
Neufeld, P., 1
Non-exclusionary conclusion, 145
Non-human animal hairs, 25, 45; *see also* Animal hairs
Non-intimate sample, *55*
Non-pigmented vellus hairs, 37
Non-scalp hairs, 63
Non-telogen hairs, 10
Normal witness, 154
Nuclear DNA (nu-DNA), 10–11, 13, 17, 23, 30, 45, 69, 74–76, 120–121, 123, 135, 143, 177, 179–180, 189, 204
Nucleic acid metabolism, 30
Null hypothesis, 129–131, *134*

O
Ogle, R., 6
Ongoing competence, 131
Opaque medulla, 104; *see also* Medulla
Oral report, 139, 140, 153; *see also* Report/reporting
Organisation of Scientific Area Committees for Forensic Science (OSAC), 171
Ottens, R., 76
Outer root sheath (ORS), 30, *33*
Ovoid bodies (OB), 95, *99–101, 103–104, 106–107*, 137, 159

P
Pangerl, E., 200, 203, 204
Paper "boat" collection kit, 51–53, *52–53*
Pelfini, C., 38
Permanent zone, 30; *see also* Upper follicle
Phaeomelanin, 14, 47
Physical anthropology, 14
Physical evidence, 53–54, 56
Physical materials, 12, 17, 27, 49–50, 54, 57, 62, 172
Pigment/pigmentation
    aggregate shapes, 95, *105*
    categories, 95
    and cell division, *35*
    and colour, 46–47
    concentration, 98
    density, *95, 96–97*
    distribution, 95
    formation, 30
    granule, 15, *80*, 95, *96–97, 101–102*
    grey, *82*
    melanin, 47
    natural-coloured, *80*
    red hair, *96*
Pili annulati, *91*
Pili torti, *89*
Pili Trianguli, *91*
Pilosebaceous unit, 30
Plate III of human hair, *5*
Positive test statements, 146
Post it Flag™, *52, 53*
Post-mortem banding (PMB), 75, *86*
Postmortem (PM), *52–53*
Post-nuclear counting, 125
Pre-germ, 27, *28*
Presidents' Council of Advisors on Science and Technology (PCAST), 2–4, 6, 12, 15, 22–23
Priestly, H., 6
Primary hairs, 37
Principles of microscopy and sample preparation, 175–176
Probabilities, estimating, 132–133
Professional judgement, 128, *129*, 131, 148, *149*
Profound colour changes (PCC), *65*

Prospective expert witnesses, 154
Protocol for hair examination,
        157–158, 161, 165, 168–169;
        see also Examinations
Proverbial towel, 20
Pubic hairs, 36, 38, *40, 44,* 54–55,
        57, *70,* 76

Q
Qualitative data, 130; *see also* Data
Quantico, 19
Quantitative data, 130

R
Ragged cuticle, 110
RCMP Hair and Fibre Section
        Methods Manual, 14, 19
Recording and recovery of hairs, 50–56
    sampling from deceased persons or
        remains, 53–54
    sampling from living persons, 55–56
    scene considerations, 50–51
    scene sampling protocols, 51–53
Recovery and collection of hairs,
        173–174
Reddish hairs, 14, 47, 77, *83,* 97, 99,
        199
Reference collection, 67
Refractive index (RI), 66
Reliability, 2–3, 15–16
Report/reporting, 139–171
    of acquired characteristics,
        191–192
    contents, 142–157
        analysis and comparison of
            material, 143–144
        cognitive bias, 152–153
        conclusions and opinions,
            144–145
        giving evidence, 153–157
        requirements, 142–143
        role of statistics and verbal
            scales for hair opinions,
            148–152
        testimony, 153–157
        wording, 145–148
    example, 158–169
    evaluative, 148

formats, 140–141
    oral reports, 140
    written reports, 140–141
    issuing of, 141
    oral, 139, 140, 153
    scope, 139
    written, *see* Written reports
Robertson, J., 7–8, 13–15, 22, 46,
        68, 93–94, 192
Roe, G. M., 126
Rogers, G., 25–27, 31–32, 37, 42–43,
        45–47
Rook, A. R., 46, 83
Roux, C., 9
Royal Canadian Mounted Police
        (RCMP), 2
Royal Statistical Society (RSS),
        148
Rudnicka, L., *88*

S
Saitoh, M., 38
Sampling
    from deceased persons or remains,
        53–54
    intimate, 55
    known, 14, 16–17, 20, 54, 62, 77,
        87, 113, 134, 136–138, 144,
        152, 193
    from living persons, 55–56
    non-intimate, 55
    scene protocols, 51–53
Scale features, 119
Scalp hair, 6, 12, 33, *34,* 35–36, 38–
        39, 42, 56, 63, 69, 76–77,
        179, 192–193, 195, *196*
Scalp regions for collecting cut hair
        samples, *58*
Scanning electron micrograph, *42*
Scanning electron microscopy (SEM),
        110
Scattered mosaic, 35
Scene considerations, 50–51
Scene sampling protocols, 51–53
Scientific analysis, 23
Scientific Working Group for
        Material Science Analysis
        (SWGMAT), 8, 171, 172

Secondary hairs, 37
Selection for DNA analysis, 72–76
Selection of hairs for nuclear
    staining, 121
Selection of hairs for nu-DNA testing,
    179–180
Seperation of human hairs, 63–69,
    173–177
Serrated cuticle, 110
Sexual assault examinations, 55
Sexual hairs, 37
Shaft diameter, 95
Shaft profile, 63, 77
Sharp-bladed implement, 196–198
Shield hairs, 63, 64
Short tandem repeat (STR), 10
Short terminal hairs, 37
Silver, A. F., 38
Sims, R. J., 38
Singed tip, 82
Smith, S., 6
Smooth cuticle, 110
Snap seal plastic bag (SSPB), 52, 54
Somatic hair shaft profiles, 39–40
Specialist hair examiner, 143
Split ends, 42, 90, 110
Split tip, 82
Spun glass hair, see Pili Trianguli
Statistical error, 133
Straight hairs, 63
Strengthen normal negative hair
    comparison conclusions, 152
Strengthen positive hair comparison
    conclusions, 152
Structure of human and animal hairs,
    176–177; see also Animal
    hairs
Supplementary report, 140
"Swiffer" pads, 53

T
Tafaro, J. T., 82
Teerink, B. J., 67
Telogen, 29–32
    follicles, 34
    growth phase, 27
    hair roots, 10
    hairs, 11

Telogen hair root nuclear staining,
    120–125
    procedure, 121–125
        apoptosis in hair, 121
        counting stained nuclei, 124–125
        DNA analysis, 125
        fixation, 124
        post-nuclear counting, 125
        selection of hairs for nuclear
            staining, 121
    reagents required, 120
    staining
        DAPI-DABCO staining, 124
        Harris's haematoxylin staining,
            124
    workplace health and safety
        (WHS), 120–121
Terminal hairs, 26, 37
Test for bleached hair, 126
Testimony and giving evidence,
    153–157
Text messages, 140
Thomas, E., 128
Thoughtful-compliance, 19
Tilstone, 1
Tip
    cut, 192, 194
    end of hairs, 47, 65, 69, 76, 97,
        110
    end of shaft, 76, 107
    rounded, 195, 203
    types, 86
Tiselius, A., 7
Training, 171–188
    competencies, 172–173
    e-platform, 171
    knowledge, 172–173
    level 1 training, 173–177
        module 2 (biology and
            chemistry of hairs), 174–175
        module 3 (principles of
            microscopy and sample
            preparation), 175–176
        module 1 (recovery and
            collection of hairs), 173–174
        module 4 (structure of human
            and animal hairs), 176–177
        overview, 173

level 2 training, 177–180
  module 1 (body area
    determination), 177–178
  module 2 (ethnic origin
    determinations), 178–179
  module 3 (selection of hairs for
    nu-DNA testing), 179–180
  overview, 177
level 3 training, 180–188
  module 1 (biology and
    chemistry of hair), 181–182
  module 2 (microscopic features
    of hair at the LPM level),
    182–183
  module 3 (microscopic features
    of hair at the TLM level),
    183–185
  module 5 (report), 187–188
  module 4 (use of comparison
    microscopy), 185–186
  overview, 180–181
scope, 171–172
Transforming data into information,
  127–131
Transient zone, 30
Translucent medulla, 104
Transmitted light comparison
  microscope, 93
Transmitted light microscopy (TLM),
  14–16, 31–32, 45, 67, 173,
  175–176, 177, 183–185, 202
Trichonodosis, 83, 91–92
Trichorrhexis invaginate, 88
Trichorrhexis nodosa, 88–92
Trichoschisis, 90
Trichothiodystrophy, 90
Triggs, B., 67
Trotter, 5

U
Ubelaker, D. H., 71
Uncertainty of measurement, 133
Unconscious bias, 92; see also Bias
Under-hairs, 63, 64
Uniserial ladder, 68
Uniseriate ladder, 119
Unstained telogen roots, 122

Upper follicle, 30; see also
  Follicles
US Department of Justice (DoJ), 3,
  12, 18
US National Research Council, 2

V
Validity as applied, 3–4, 9, 23
Vellus hairs, 28, 37
Verbal qualifiers, 150–151
Visual assessment of natural hair
  colour, 46–47
Visual literacy, 128, 137
Voir dire situation, 156

W
Watkins, I., 150
Wavy hairs, 63, 64
Weaken normal negative hair
  comparison conclusions, 152
Weaken positive hair comparison
  conclusions, 151
Wickenheiser, R. A., 6
Wide aeriform lattice, 68
Wide medulla lattice, 68
Windscreen impact, 193–195, 196
Witnesses
  adequate manner as, 157
  expert, 153–157
  normal, 154
  personal demeanour as, 157
  prospective expert, 154
Wittig, M., 8
Wortmann, F. J., 69
Written reports, 139–141; see also
  Report/reporting
  certificates, 141
  e-mails, 140
  full report, 141
  interim report, 140
  laboratory report, 141
  requirements of, 142
  supplementary report, 140
  text messages, 140

Y
Yellow hair, 14, 47, 77, 80, 85